T0327586

CHARACTERISTIC MODES

CHARACTERISTIC MODES

Theory and Applications in Antenna Engineering

YIKAI CHEN AND CHAO-FU WANG

Temasek Laboratories
National University of Singapore

Copyright © 2015 by John Wiley & Sons, Inc. All rights reserved

Published by John Wiley & Sons, Inc., Hoboken, New Jersey
Published simultaneously in Canada

No part of this publication may be reproduced, stored in a retrieval system, or transmitted in any form
or by any means, electronic, mechanical, photocopying, recording, scanning, or otherwise, except as
permitted under Section 107 or 108 of the 1976 United States Copyright Act, without either the prior
written permission of the Publisher, or authorization through payment of the appropriate per-copy fee to
the Copyright Clearance Center, Inc., 222 Rosewood Drive, Danvers, MA 01923, (978) 750-8400, fax
(978) 750-4470, or on the web at www.copyright.com. Requests to the Publisher for permission should
be addressed to the Permissions Department, John Wiley & Sons, Inc., 111 River Street, Hoboken,
NJ 07030, (201) 748-6011, fax (201) 748-6008, or online at http://www.wiley.com/go/permissions.

Limit of Liability/Disclaimer of Warranty: While the publisher and author have used their best
efforts in preparing this book, they make no representations or warranties with respect to the accuracy
or completeness of the contents of this book and specifically disclaim any implied warranties of
merchantability or fitness for a particular purpose. No warranty may be created or extended by sales
representatives or written sales materials. The advice and strategies contained herein may not be
suitable for your situation. You should consult with a professional where appropriate. Neither the
publisher nor author shall be liable for any loss of profit or any other commercial damages, including
but not limited to special, incidental, consequential, or other damages.

For general information on our other products and services or for technical support, please contact our
Customer Care Department within the United States at (800) 762-2974, outside the United States at
(317) 572-3993 or fax (317) 572-4002.

Wiley also publishes its books in a variety of electronic formats. Some content that appears in print may
not be available in electronic formats. For more information about Wiley products, visit our web site at
www.wiley.com.

Library of Congress Cataloging-in-Publication Data:

Chen, Yikai, 1984–
Characteristics modes : theory and applications in antenna engineering /
 Yikai Chen, Chao-Fu Wang, Temasek Laboratories, National University of Singapore.
  pages cm
 Includes bibliographical references and index.
 ISBN 978-1-119-03842-9 (hardback)
1. Antennas (Electronics) I. Wang, Chao-Fu, 1964– II. Title.
 TK7871.6.C457 2015
 621.382'4–dc23

2015002781

Set in 10/12pt Times by SPi Global, Pondicherry, India

Printed in the United States of America

10 9 8 7 6 5 4 3 2 1

1   2015

CONTENTS

LIST OF FIGURES

LIST OF TABLES

PREFACE

Characteristic mode (CM) theory has received a great deal of attention in the field of antenna engineering in recent years. It is increasingly becoming a popular and general approach to characterize the modal resonant behavior of arbitrarily shaped objects such as antennas and scatterers. Its resultant characteristic currents form a weighted orthogonal set over the surface of the objects and their corresponding characteristic fields form an orthogonal set over the radiation sphere at infinity. These attractive features are useful in terms of the designs of reconfigurable antenna systems. Fully exploiting and making use of the characteristic modes of an antenna can significantly enhance its fundamental parameters, such as efficiency, gain, polarization purity, compactness, as well as much flexibility in the design of its excitation structures.

Quite a number of papers on the CM theory and its applications have been published by different research groups in many journals and conference proceedings. Although these papers are readily available to be read for understanding the CM theory through many electronic databases, some fundamental questions are still frequently raised by antenna engineers in many occasions, formal or informal. For instance, the top five questions raised are as follows:

1. What are the characteristic modes?
2. When do we need the CM theory for antenna designs?
3. What kind of antenna structures or materials can be analyzed using CM theory?
4. How to apply CM theory in practical antenna designs?
5. What are the merits of the characteristic modes in antenna designs?

Therefore, it is now the right time to write a book that can act as a one-stop reference to antenna scientists and engineers. We believe that this book will make CM theory become a standard approach for antenna analysis and design. We hope this book will stimulate more novel antenna designs based on the CM theory. We also hope this book can help eliminate some misconceptions on antenna design after the underlying physics of antenna structures are understood from the viewpoint of characteristic modes. For example, antenna design is often regarded as an easy task that can be completed effortlessly with the help of user-friendly electromagnetic simulators equipped with capable optimizers. With this misconception, it is claimed that any kind of antennas can be produced upon the request of sponsors despite the lack of practical design experience. However, this cognition is not true as the product of computer-aided design tools lacks physical meanings and can never replace the wisdom of human beings. Based on the authors' personal experience in CM-based antenna developments, the antenna design in turn is a scientific, rigorous, and systematic process. It should start from certain required radiation performance and end with practical and beautiful antenna structure, which is physically determined by the required radiation performance and produces better or optimal radiation performance if possible.

Technically, this book aims to provide antenna scientists and engineers with up-to-date knowledge on the CM theories, numerical implementations, as well as novel antenna design concepts, methodologies, and typical antenna designs developed based on the CM theory. It gives a full picture of the CM theory family, ranging from the CM theory for perfect electric conductors, planar antennas in multilayered medium, dielectric material bodies, and multiport networks. This book is the first book that clearly describes how to implement the various CM theories into the designs of practical antenna systems in a systematic manner. It should be valuable for antenna engineers and designers to seek new solutions for the antenna design problems they face, which are too difficult to solve conventionally. It is also an ideal book for antenna scientists attempting to acquire in-depth physical insights in the radiation mechanisms of many conventional antennas.

Moreover, this book includes authors' plentiful contributions featured on the study of the CM theory for solving practical problems. A wide range of applications including electrically small antennas, microstrip patch antennas, dielectric resonator antennas, multiport antenna systems, antenna arrays, and platform-integrated antenna systems are also covered. Through these illustrative design examples, readers will discover the great potential of the CM theory in many challenging antenna designs. The following topics will be covered in the six chapters of this book:

- A detailed review of the CM theory and its applications in radiation and scattering problems;
- Comprehensive descriptions for the CM theories of various electromagnetic structures;
- Numerical algorithms of the CM theories and their implementations;
- CM-based antenna design concepts and methodologies of various antenna system designs.

Reading of this book may require the basic knowledge of electromagnetic theory, computational electromagnetics, and antenna theory and technology. As a book for advanced topics in antenna engineering, we also assume the readers are familiar with various commonly used antennas in the industry, such as the microstrip patch antennas, dielectric resonator antennas, and other electrically small antennas in wireless communications.

Although we have prepared the manuscript of this book with great care, typos and errors will inevitably occur. We would greatly appreciate the notice of them via e-mail to Yikai Chen at ykchen@ieee.org and Chao-Fu Wang at CFWang@nus.edu. sg or @ieee.org. Any comments, suggestions, and constructive criticisms will be also most welcome and may be forwarded to our e-mail addresses.

There are many people that we would like to thank who directly or indirectly contributed to the success of this book. We are indebted to many pioneering scholars and colleagues who have made great contributions to the study of the CM theory. Their great work has provided us a lot of fundamental materials making this book possible. We are happy to work with our colleagues in the Temasek Laboratories at National University of Singapore (TL@NUS), from whom we have received and learnt a lot. We would like to particularly thank Professor Hock Lim, Mr. Joseph Ting Sing Kwong, and Professor Boo Cheong Khoo for their great support and encouragement, Dr. Tan Huat Chio and Dr. Fu-Gang Hu for their very useful discussions, Mr. Peng Khiang Tan for his assistance in the fabrications and measurements of antennas described in this book.

Chao-Fu Wang would like to particularly thank Professor Jian-Ming Jin of University of Illinois at Urbana-Champaign, Professor Da-Gang Fang of Nanjing University of Science and Technology, and Professor Weng Cho Chew of University of Illinois at Urbana-Champaign, who have taught him a lot about electromagnetics and helped him a lot through many occasions. Thanks are expressed particularly to Professor Joshua Le-Wei Li of Monash University, Professor Qing Huo Liu of Duke University, Professor Yang Hao of Queen Mary University of London, Zhizhang (David) Chen of Dalhousie University, and Professor Ji Chen of University of Houston, for their great help and support during the writing of this book. Chao-Fu Wang would also like to thank Ms. Xin-Xin Wang for her help in wording some of the chapters of this book. Yikai Chen would also like to take this opportunity to thank Professor Shiwen Yang, Professor Zaiping Nie, and Professor Jun Hu of University of Electronic Science and Technology of China (UESTC), for their encouragement and patience when he pursued the Ph.D. degree in electromagnetics at the UESTC.

We are really grateful to the staff of John Wiley & Sons, Inc., especially Editor Brett Kurzman and Editorial Assistant Alex Castro of Global Research, Professional Practice and Learning, for their interest, support, and cooperation. Last but not least, we are grateful to our wives for their patience and loving support.

Temasek Laboratories   YIKAI CHEN AND

National University of Singapore   CHAO-FU WANG

1

INTRODUCTION

1.1 BACKGROUNDS

Over the past few decades, the field of antenna engineering has undergone significant progress. Many new techniques and design concepts have been developed to overcome a myriad of challenges experienced in antenna engineering. Amongst the advancements, the increasing characteristic mode (CM) theory study, focusing on its extensive implementations in many critical antenna designs, is one of the exciting breakthroughs in antenna engineering. Its promising potentiality has been constantly attracting the attention of antenna engineers. The CM theory and its applications in antenna engineering are the topics of this book.

The booming of wireless communication is an important driving force for the advancement of antenna technology. Antennas are the sensors of wireless communication systems. They find wide range of applications from terminal devices (such as mobile phones) to advanced communication systems on aircrafts, ships, and so on. Antennas transfer microwave energy from transmission system to propagating waves in free-space and vice versa. Strong demands like small physical size, low weight, low cost, wideband/multiband bandwidth, reconfigurable capabilities, or even aesthetic consideration are increasingly specified as a must in modern antenna designs. The inherent challenges in these demands thus have further propelled the advance of antenna technology.

The rapid growth of numerical electromagnetic (EM) modeling techniques plays another vital role in antenna technology advancement. The numerical EM modeling

Characteristic Modes: Theory and Applications in Antenna Engineering, First Edition.
Yikai Chen and Chao-Fu Wang.
© 2015 John Wiley & Sons, Inc. Published 2015 by John Wiley & Sons, Inc.

techniques can provide an accurate way to validate antenna performance before carrying out expensive fabrications and measurements. Consequently, numerous in-house or commercial software packages based on the method of moments (MoM) [1], the finite element method (FEM) [2, 3], and the finite difference time domain (FDTD) method [4–6] are extensively used in antenna designs. Given the antenna geometry and excitation structure, numerical techniques are able to simulate any antenna parameters.

However, from the practical point of view of antenna design problem, these numerical EM modeling techniques provide little information on the physical aspects of an antenna to be designed. The lack of physical insights brings difficulties in the further optimization of the antenna structure and feedings for achieving enhanced radiation performance. Therefore, antenna designs are heavily reliant on the designer's experience and knowledge. In the worst cases, antenna designs become a trivial task where antenna engineers blindly modify the antenna and feeding structures and simulate the antenna performance via numerical EM modeling tools iteratively.

For the sake of convenience, numerical EM modeling techniques are extensively combined with modern evolution optimization algorithms such as the genetic algorithm (GA) [7] to help mitigate the heavy workload in antenna tunings. The assumption is that the optimization algorithm will eventually arrive at the expected antenna performance after the exhaustive search in their decision space. However, this is not always true in all antenna design problems. More often than not, the automatic optimization algorithm returns to a complicated antenna structure with the satisfactory level of performances. However, the complexity of the resultant antenna structure makes it too hard to understand the underlying radiation mechanism. In this case, the design will be generally regarded as a lack of scientific knowledge. Therefore, overdependence on such brute force techniques is not a good way in antenna research. At least, it should not become antenna engineers' primary choice.

It is evident that a successful antenna design is highly dependent on previous experiences and the physical understanding of antennas. To grasp such knowledge may require many years of practical exercise. The experience, however, is hard to be imparted from one to another, as such personal experience is usually formed based on one's understanding of conventional antenna design concept introduced in textbooks. With such experience, solutions to some critical antenna design problems (e.g., the problems in Chapter 6) are usually not available.

Based on authors' personal understanding, an ideal antenna design methodology should allow one to achieve optimal antenna performance in a systematic synthesis approach with very clear physical understandings. However, such antenna design methodology does not exist until the antenna community recognizes the great potential of the CM theory in antenna engineering. In the past decade, the extensive applications of the CM theory in antenna designs have witnessed the roadmap of the development of this ideal antenna design methodology.

In the new millennium, studies on the CM theory have revealed its promising potential in a variety of antenna designs. The CM theory makes antenna design much easier than ever as antenna engineers need not depend heavily on personal experiences or brute force optimization algorithms. Meanwhile, the CM theory provides an easy way to understand the physics behind many key performances such as the bandwidth, polarization, and main beam directions. These physical understandings provide a greater degree of freedom in terms of design. As compared to traditional antennas, the antennas designed with the CM theory have more attractive electrical performances and configurations. Based on the recent advances made by the authors from the Temasek Laboratories at National University of Singapore (TL@NUS), this book discusses the CM theory and the CM-based design methodologies for a wide range of antenna designs.

1.2 AN INTRODUCTION TO CHARACTERISTIC MODE THEORY

The CM theory is a relatively new topic in the antenna community. Its great potential in antenna engineering has not been widely recognized till 2000. In the following subsections, several well-known modal methods for the analysis of particular antenna problems are being reviewed first providing preliminary knowledge about what modal analysis is about. It also illustrates how these modal methods help in practical antenna design. Next, we address some fundamental questions: why the characteristic modes were proposed and what are characteristic modes? Furthermore, the primary features of the CMs would be discussed. Finally, we briefly introduce CM variants that are distinct in terms of structures and materials to meet different radiation/scattering requirements. Detailed formulations and applications of these CM variants are discussed through Chapters 2–6.

1.2.1 Traditional Modal Analysis in Antenna Engineering

There is a long history in the development of modal analysis methods for a variety of problems in electromagnetics. The most famous one would be the modal expansion technique for infinite long waveguides [8–11]. This modal expansion technique handles the electromagnetic field inside a closed structure. It calculates the cut-off frequency and modal field pattern of each propagating mode inside the waveguides. These modes are primarily determined by the boundary conditions enforced by the cross-section of the waveguide. However, it is exceptionally difficult to implement modal analysis for open problems (radiation or scattering). Thus if the resonant behavior and radiation performance of an antenna can be interpreted in terms of the mode concept, great convenience will be brought to the analysis, understanding, and design of antennas.

In antenna engineering, there are three well-known modal analysis methods for particular antenna structures, namely, spherical mode, cavity model, and dielectric waveguide model (DWM). They will be briefly reviewed to show their attractive features and primary limitations.

1.2.1.1 Spherical Mode The first modal analysis method for antenna problems was the spherical mode method proposed in 1948 [12]. It discussed the physical limitations of the omnidirectional antennas in terms of antenna quality (Q) factor. It is now known as the Chu's Q criterion. Later in 1960, Harrington used the spherical mode analysis to discuss the fundamental limitations with respect to the gain, bandwidth, and efficiency of an antenna [13]. These pioneering works were based on the assumption where the radiating fields can be expanded using the spherical modes within a sphere that completely encloses the antenna. Therefore, the radiated power can be calculated from the propagating mode within the sphere. The limitations of an electrically small antenna in terms of the Q factor can be determined from the size of the antenna. However, as the size of the antenna increases, it gives rise to many propagating modes. In order to take into account all the propagating modes, the enclosing sphere has to be large enough. On the other hand, it is difficult to compute the modal coefficients for all the propagating modes in the large sphere. Thus, in general, the spherical mode method is limited to antennas with very small electrically size.

1.2.1.2 Cavity Model   The cavity model was introduced by Lo et al. in 1979. It is a well-known modal analysis method for microstrip antennas [14], where the microstrip antenna is modeled as a lossy resonant cavity. As shown in Figure 1.1, the cavity model is set up as a region bounded by electric walls on both of the top and bottom interfaces, and magnetic walls along the perimeter of the patch antenna. In general, the substrate is assumed to be very thin in terms of thickness. The thin substrate thickness ensures the suppression of the surface wave. It also ensures that the field inside the cavity is uniform in the direction of the substrate thickness [14, 15]. Meanwhile, the cavity model assumes that the fields underneath the patch are expressed as the summation of various resonant modes. Moreover, the cavity model takes into account the fringing fields along the perimeter by extending a small offset distance out of the patch periphery. The far-fields are computed from the equivalent magnetic currents around the periphery. The cavity model also accounts for the higher order resonant modes and the feed inductance. It offers a simple yet clear physical insight into the resonant behavior of microstrip antennas. However, it is only applicable to regular patch shapes such as the rectangular, circular, elliptical, and triangular patches. The computation of resonant modes for arbitrary shaped patches using cavity model is a challenging task. In addition, the cavity model is also not suitable for the analysis of multilayered or thick substrate microstrip antennas.

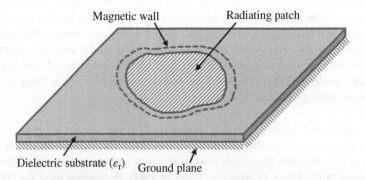

FIGURE 1.1   The cavity model of a microstrip antenna.

1.2.1.3 Dielectric Waveguide Model The DWM is another well-known modal analysis method [16], which was developed to evaluate the resonant frequencies of rectangular dielectric resonator antenna (DRA). The DWM is generally accurate enough and has been widely adopted in practical rectangular DRA designs [17–19].

In the DWM, the rectangular DRA is modeled as a truncated dielectric waveguide. Referring to the rectangular DRA sitting in the coordinate system as shown in Figure 1.2, where $a > b > d$ the first three lowest modes are TE_{111}^{z}, TE_{111}^{y}, and TE_{111}^{x}, respectively. As the analysis method for all of these modes is similar, only the resonant frequency determination for the TE_{pqr}^{y} mode is being discussed and provided below. The characteristic equations for the wavenumber k_x, k_y, and k_z by using the Marcatili's approximation technique [17] are as follows:

$$k_x a = p\pi - 2\tan^{-1}(k_x / \varepsilon_r / k_{x0}), \quad p = 1,2,3\ldots$$
$$k_{x0} = \left[(\varepsilon_r - 1)k_0^2 - k_x^2\right]^{1/2} \tag{1.1}$$

$$k_y b = q\pi - 2\tan^{-1}(k_y / k_{y0}), \quad q = 1,2,3\ldots$$
$$k_{y0} = \left[(\varepsilon_r - 1)k_0^2 - k_y^2\right]^{1/2} \tag{1.2}$$

$$k_z d = r\pi - 2\tan^{-1}(k_z / \varepsilon_r / k_{z0}), \quad r = 1,2,3\ldots$$
$$k_{z0} = \left[(\varepsilon_r - 1)k_0^2 - k_z^2\right]^{1/2} \tag{1.3}$$

where k_{x0}, k_{y0}, and k_{z0} are the decay constants of the field outside the DRA, and k_0 is the free-space wavenumber. The wavenumbers k_x, k_y, k_z, and k_0 satisfy the following formulation:

$$k_x^2 + k_y^2 + k_z^2 = \varepsilon_r k_0^2 \tag{1.4}$$

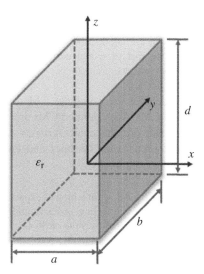

FIGURE 1.2 An isolated rectangular dielectric resonator antenna.

By solving the characteristic equations simultaneously, one can determine the resonant frequencies of the TE^y_{pqr} modes.

However, the DWM is only applicable to rectangular DRAs. For cylindrical DRAs and other DRAs with more complicated shapes, one would usually resort to some other engineering formulas [20] or full-wave simulations.

In summary, the modal analysis methods provide clear physical insights to the resonant behavior and modal radiation performance. It is understood that each of these modal analysis methods is theoretically developed for solving a particular kind of problems, thus they are not versatile for general antenna problems. Therefore, it is essential to develop more general modal theory for a variety of antenna problems. The CM theory was initially developed as a versatile modal analysis tool for antennas with arbitrary structures and materials. The CMs together with its metric parameters explicitly give the following useful information for antenna analysis and design:

- Resonant frequencies of dominant mode and high order modes;
- Modal radiation fields in the far-field range;
- Modal currents on the surface of the analyzed structure;
- Significances of the modes at a given frequency.

1.2.2   Definition of Characteristic Modes

Garbacz proposed the CM theory for the first time in 1965 [21]. At that time, the definition for CMs was not explicitly given. Instead, an assumption for the CMs was given before addressing their numerical computations:

> The basic assumption is made that the scattering or radiation pattern of any object is a linear combination of modal patterns, characterized by its shape, and excited to various degrees at the terminals (when the object acts as a radiator) or by an incident field (when the object acts as a scatterer). The current distribution over the object is assumed to be decomposed into an infinite number of modal currents, each of which radiates a characteristic modal pattern independent of all others.

Garbacz and Turpin [22] addressed the complete CM theory and the expansion method for the computation of CMs. They defined CMs as a particular set of surface currents and radiated fields that are the characteristics of the obstacle and are independent of any external source. Later, Garbacz and Pozar [23, p. 340] gave a definition for CMs:

> Characteristic modes form a useful basis set in which to expand the currents and fields scattered or radiated by a perfectly conducting obstacle under harmonic excitation. They possess orthogonality properties both over the obstacle surface and the enclosing sphere at infinity; they succinctly relate the scattering operator and impedance operator representations of the obstacle; and they exhibit interesting mathematical properties which may be interpreted physically in terms of radiated and net stored powers.

The definitions for the characteristic modes show that the characteristic modes constitute to a very special orthogonal set in the expansion of any possible induced currents on the surface of the obstacle. Moreover, the far-fields associated with these orthogonal currents possess orthogonality properties over the radiation sphere in the infinity.

1.2.3 Primary Properties of Characteristic Modes

The CM theory was initially proposed for perfectly electrically conducting (PEC) bodies. For the PEC case, there are two major features that make it attractive in antenna engineering and scattering problems [23]:

- Any PEC bodies with surface S is associated with an infinite set of real characteristic currents $\mathbf{J}_n$ on S. Each of $\mathbf{J}_n$ radiates the characteristic electric field $\mathbf{E}_n$ into the free space. The tangential component of $\mathbf{E}_n$ on S is equiphase. Associated with each CM, there is a characteristic angle $\alpha_n \in [90°, 270°]$ defining the phase lag between $\mathbf{J}_n$ and $\mathbf{E}_n^{\tan}$. The characteristic angle is an important metric parameter in the CM theory, which indicates the resonant behavior and the energy storage of each mode.
- Due to the orthogonality of $\mathbf{J}_n$ over S as well as the orthogonality of $\mathbf{E}_n$ over the radiation sphere at infinity, the CMs form a useful basis set in the expansion of any possible currents or fields associated with the perfectly conducting body.

In addition to the characteristic angle, other metric parameters were often used to describe the resonant behavior of each mode. Each of them has its unique features to illustrate the physics of each mode. These parameters and their physical interpretations will be presented in Chapter 2.

1.2.4 Variants of Characteristic Modes

There are many variants of CMs for the analysis of structures with varying configurations and materials. Among these variants, the initial work done by Garbacz has to be highlighted first [21]. Garbacz stated that arbitrary shaped PEC obstacles possessed CMs. These CMs are only dependent on the shape and size of the PEC obstacles. They are independent on any specific excitation or sources. Most importantly, these modes can be used to expand any possible currents and scattering fields due to the PEC obstacle. He further demonstrated that the CMs of a PEC sphere were identical to the spherical wave functions, and the CMs of an infinite circular PEC cylinder were identical to the associated cylindrical wave functions. This early observation showed that modal analysis for arbitrary-shaped PEC obstacles can be achieved through the implementation of CM analysis.

This initial work motivated Garbacz and Turbin to develop an approach to compute the CMs for arbitrary obstacles. In Ref. [22], they derived a generalized scattering matrix by matching the external and internal fields to the obstacle at sampling points. A characteristic equation was then formulated by diagonalizing this scattering matrix. By solving the characteristic equation, the CMs of the obstacle were obtained. Although this approach required substantial computations in getting the characteristic equation, it was the first successful attempt to compute the CMs for arbitrary PEC bodies.

In 1971, R. F. Harrington and J. R. Mautz reformulated the CM theory based on the operator in the electric field integral equation (EFIE) for PEC objects [24, 25]. This operator relates the surface current to the tangential electric field on the surface of PEC body. By diagonalizing the operator and choosing a particular weighted eigenvalue equation, they obtained the same modes as defined by Garbacz. Harrington's approach produced explicit and convenient formulas for the computation of CM currents and fields. More often than not, the CM formulation and computation implemented in recent years are primarily based on Harrington's approach.

An EM structure with many input/output ports is commonly treated as a multiport network system. Accurate analysis of the transmission and scattering properties of a multiport network system can be carried out via the port impedance matrix [Z], the port admittance matrix [Y], and the scattering matrix [S]. With these port parameter matrixes, similar to the CMs for PEC bodies [24], Harrington and Mautz developed the characteristic port modes for the eigen analysis of an N-port network [26]. The characteristic port modes were computed from the mutual impedance matrix of an N-port network system. They formed a convenient basis set for expressing the field scattered or radiated from an N-port network. It should be noted that the characteristic port modes were actually defined in a manner analogous to those for continuously loaded bodies. With the help of such kind of analogue in representing functionality, the analysis, synthesis, and optimization of the N-port systems (e.g., antenna array) become conceptually simpler. Many researchers have applied the characteristic port modes in the analysis and design of multiport antenna systems such as the multiple input multiple output (MIMO) antenna systems [27] and the wideband antenna arrays [28].

Both dielectric and magnetic materials are widely used in radiation systems. Therefore, CM theory for structures with different materials is highly demanded. In the 1970s, Harrington and his colleagues attempted to extend the CM theory to structures involving different materials [29, 30]. However, the correctness of their formulations had not been well demonstrated in solving practical problems until 2013. Several new investigations to these earlier formulations have been addressed in the recent years [31–35]. With regards to this CM variant, two new CM variants are respectively developed for the CM analysis of planar patch antennas in multilayered dielectric medium and for the CM analysis of dielectric bodies. The latter CM formulation offers useful applications in the analysis and design of DRAs. The theory and applications of these two CM variants in practical antenna designs will be discussed in depth in Chapters 3 and 4, respectively.

In many cases, the far-fields on the entire radiation sphere are often not necessary. For instance, backward radiations from a directive antenna or antenna array, backed by a large ground plane, are usually very small. It may not be necessary to consider the backward radiations in the design stage if only the radiation characteristics in the upper half-space are of interest. In such circumstance, if the far-fields derived from the CM theory can be orthogonal to each other over an interested section (say, the interested upper half-space) on the radiation sphere, such an orthogonality will bring great convenience in the analysis and design of many antennas and antenna arrays. Inagaki and Garbacz [36] developed a new CM formulation to address this issue, which is generally known as Inagaki modes. The Inagaki modes are more generalized than the conventional CMs and the orthogonal mode fields can be imposed on any section of the radiation sphere at infinity. Afterward, a modified version of the Inagaki modes was proposed by Liu [37]. This modified Inagaki modes were often referred to as the generalized CMs that provided advantages in terms of versatility and computational efficiency in the computation of the Inagaki modes.

Apertures and slots are another kind of EM structures that are widely used for radiation purpose. The aperture problems have been investigated by many researches, even back to several decades ago [38]. Small apertures in an infinite conducting plane are usually treated using the Bethe-hole theory [39]. If the small aperture is in a nonplanar surface, the Bethe-hole theory is usually used as an approximation. When the aperture increases to several wavelengths, one usually resorts to the solution of an appropriate integral equation. To solve the aperture problems with arbitrary size and shape in a conducting plane, a CM theory for various aperture problems was developed by Kabalan [40–63]. Specifically, Kabalan developed the CM theory for the following aperture problems:

- Slots in a conducting plane [40, 41].
- Rectangular aperture in a conducting plane [49].
- Slot in a conducting plane separating different media [42].
- Slots in a conducting cylinder [43, 44].
- Dielectric-filled conducting cylinder with longitudinal slot [45].
- Multiple slots in a conducting plane [46].
- Aperture-fed waveguide problem [47].
- Dielectric conducting cylinder with multiple apertures [48].
- Two-half space regions separated by multiple slot-perforated conducting planes [50].
- Parallel plate-fed slot antennas [51].

In all of these aperture CM formulations, the modes have the same attractive properties as those in the PEC objects:

- The characteristic magnetic currents are real.
- The characteristic magnetic currents are weighted orthogonal over the aperture region.
- The characteristic fields are orthogonal over the infinite radiation sphere.

1.3   CHARACTERISTIC MODES IN ANTENNA ENGINEERING

The important features of CMs have given rise to a variety of applications in antenna engineering. In this section, the CM-related antenna topics, since it was proposed in 1965 [21], are reviewed. We hope that this section will leave a deep impression on the readers where the CM theory is an efficient antenna design methodology with very clear physical insights for many challenging antenna designs.

Figure 1.3 shows the number of publications on CMs since 1965. The statistical data comes from the well-known Scopus database. It can be seen that CMs were not so popular in the first 35 years. Most of the earlier CM researches were carried out at the Syracuse University and the Ohio State University, USA. However, in the new millennium, the number of publications on CM has been gradually increasing since 2000 and a new peak was achieved in the year of 2014. This may be due to the fact that the CM approach is an attractive tool in the design of multiple antenna systems on the chassis of mobile handset devices. The rapid growth of smart phones has strongly propelled the technology advancement of multiple antenna system designs within a very crowded space. Studies have shown that the CM approach is the most promising technology to address such critical antenna designs and this trend has become more evident after the year of 2006.

Figure 1.4 shows the research topics related to CM theory and their publication percentages after the year of 1980. Thirty-six percent of publications reported about the mobile handset antenna designs using CMs. Together with Figure 1.3, it is evident that large number of CM research was conducted on the CM applications to mobile handset antenna designs since 2006. The second hottest topic after that would be the CM formulation developments for apertures and slots. As discussed in the preceding subsection, these developments were carried out by Kabalan and his colleagues at the American University of Beirut from 1987 to 2004 [40–63]. It is expected that these fruitful CM formulations will discover great applications to the analysis of transmission, reflection, coupling, and scattering in aperture problems.

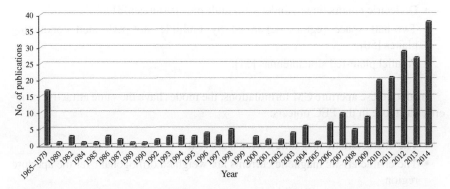

FIGURE 1.3   Yearly publications on characteristic modes. The statistical data comes from the Scopus database [64].

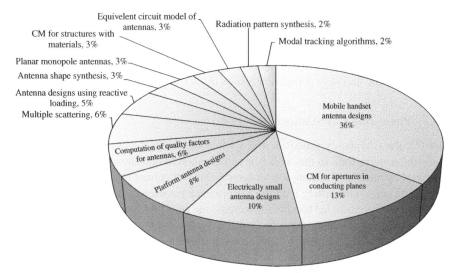

FIGURE 1.4 CM research topics and their publication percentages after the year 1980.

The topic of platform antenna designs using CMs ranks third among various CM topics. When CM theory found its applications in platform antenna designs in the late 1970s [65], the structural antenna concept using CM was also formed at the time of this pioneering study. In this approach, the existing platform was excited and used as the radiator, and the CMs of the platform were exploited to control the currents on the platform for designated radiation performance. Later, CM theory was extensively adopted in various platform antenna designs, especially in low-frequency bands such as the HF, VHF, and UHF.

Due to the vast number of publications on CM-based antenna topics, it is impossible to include a full collection of all the CM researches in this chapter. Therefore, with reference to Figure 1.4, the CM-related topics would be categorized into the following 11 subjects:

1. Mobile handset antenna designs
2. Electrically small antenna designs
3. Platform antenna designs
4. Computation of quality factors for antennas
5. Antenna designs using reactive loading
6. Antenna shape synthesis
7. Planar monopole antennas
8. CM for structures with dielectric materials
9. Equivalent circuit model of antennas
10. Radiation pattern synthesis and decomposition
11. Modal tracking algorithms

The following subsections will give a comprehensive introduction to each of the listed subjects. It is presented according to the reporting time of the CM research works, that is, carried out before 1990 and after 1990. Readers are encouraged to consult the references for more detailed explanation on the theory, methodology, implementation, and/or experimental results in each of the subject.

1.3.1 Pioneering CM Studies (1965–1990)

In 1972, the CM theory found its first application in the synthesis of desired radar scattering patterns [66]. In this work, the authors showed that any real current could be resonated by reactively loading the scatterers. This real current may be chosen to meet the specifications concerning required performance parameters, such as minimize the quality factor for enhanced bandwidth, increase the gain, and synthesize a desired radiation pattern. In particular, they developed a synthesis procedure to obtain the real current such that the far-field pattern approximates the desired one in a least mean-square sense. This initial concept laid the foundation for further radiation pattern or scattering pattern synthesis using the CM theory. It also pointed out that these reactive loads could be realized in the form of tuned slots and stubs on the scatterers.

In 1973, Yee and Garbacz developed simple quadratic forms for the self- and mutual-admittances of delta-gaps located along a perfectly conducting thin wire [67]. They demonstrated that once the characteristic currents and eigenvalues had been resolved, it was easy to calculate the admittances for any number of gaps located on the wires. The developed quadratic forms were useful for small array design and analysis, which formed a logical basis for decoupling the gaps for possible radiation pattern control. These quadratic forms were also useful in some antenna array optimizations. This was the first time that the CM theory was applied in the analysis of antennas.

In 1974, Harrington and Mautz developed a procedure to maximize the radar cross-section by reactively loading an N-port scatterer at the current maximum [68]. In this work, they first synthesized the real port currents that would maximize the radiation power gain. Then, reactive loads for the resonating of these real port currents were computed using the characteristic port modes. This work enlightened the applications of the characteristic port modes in antenna array designs. We will address the technical details of this topic in Chapter 5.

Later, Newman proposed a very attractive structural antenna concept for platform-integrated antenna designs in 1977 [65]. He demonstrated that the efficiency of a small antenna could be substantially increased through properly locating it on its supporting structure. CMs of the supporting structure were used to determine the optimum locations of such small antenna. He highlighted this new concept in Ref. [65], where the small antenna does not function as the primary radiator, but rather as a probe to excite the currents on the supporting structure. Since the supporting structure is not always electrically small, it can be an effective radiator. This was the first paper reported that the CMs of an antenna platform would dominate the radiation performance. The CM-based structural antenna concept was then successfully applied in the design of vehicle mounted sky-wave communication antennas [69, 70]. The vehicle platform and its installed loop antenna were treated as an entire radiating structure. CM analysis was then performed for this platform-loop integrated structure. The appropriate excitation and control of key modes was attained to realize the synthesis of the desired radiation patterns and significant performance enhancement over those from conventional whip antennas. It was illustrated that the platform CMs offered many design freedoms in terms of controlling of radiation patterns, enhancing gain and efficiency, as well as the optimal feeding placement. More technique details will be covered in Chapter 6.

To obtain a designated far-field radiation pattern from an unknown antenna structure, Garbacz and Pozar proposed an approach to synthesize the shape of the antenna such that the far fields of its dominant CM approximate the designated one [23]. With the synthesized antenna shape, one would only need to excite the dominant CM and suppress the unwanted modes. They also investigated many practical feedings to effectively excite the desired modes while ineffectively exciting high-order undesired modes. This investigation paved the way for the feeding designs for the excitation of platform CMs.

In 1985, Harrington and Mautz reported another important contribution to the CM theory [71]; the CM theory for problems consisting of two regions coupled by an aperture of arbitrary size and shape was developed. It created a new area of study for the CMs. Later, Kabalan's research group followed the work reported in Ref. [71] and reported a complete CM theory for various aperture problems [40–63]. Readers can refer to the preceding sections for more details of Kabalan's studies on aperture CMs.

The above work was carried out before 1990. Major applications of these findings include platform antenna designs, radar scattering controlling, radiation pattern synthesis, and antenna shape synthesis for designated radiation pattern. Theoretical investigations to aperture problems and electromagnetic structures with dielectric materials were also carried out in the earlier times. We shall see, in the next subsection, how the major application of the CM theory turns to the mobile handset antenna designs. The resurgence of the CM theory in antenna engineering is indeed due to the increasing demands on personal mobile communication devises.

1.3.2 Recent CM Developments (1991–2014)

This section describes the recent CM developments with regard to the 11 subjects as mentioned in the preceding sections. The challenges, CM-based techniques, and major achievements of each subject are presented in the following.

1.3.2.1 Mobile Handset Antenna Designs In the past decade, CM theory found the widest applications in the developments of mobile handset antenna designs. A large number of publications have reported on the mobile handset antenna designs using the CM theory [72–136]. Evidently, the CM theory has become a standard approach in these developments. Based on the CMs of the chassis of the mobile handset devices, the antenna community has made the following achievements:

- Bandwidth enhancement of mobile handset antennas: If many efficient radiating modes are distributed evenly within a frequency band, all of these modes can be excited at the same time to achieve wide bandwidth. However, the bandwidth of each chassis mode is usually quite narrow. Thus, many techniques were proposed to generate a set of efficient radiating CMs along a wide frequency band. Typical designs included adding a bezel above the chassis [72–74] or introducing a strip along the length of the chassis [75–77]. To achieve wideband handset multiple antenna system designs, capacitive coupling element and inductive coupling elements were studied [78–82].

- Port decoupling in handset MIMO antenna designs: Due to the crowding space in handset MIMO antennas, there are usually strong mutual coupling among multiple ports. The orthogonality of CMs provides an efficient way to decouple the couplings, even if the ports are physically closed with each other [83].

- Reconfigurable handset antenna designs: As the characteristic fields of each mode are orthogonal with each other, different modes feature different characteristic far-field patterns. The diversity of these characteristic fields could be exploited for reconfigurable handset antennas designs [84–86].

- Radiation efficiency enhancement: By efficiently exciting the dominant CM of the chassis, the chassis becomes the radiator. On the other hand, the size of the chassis of a handset device is usually comparable with the wavelength. Due to the large size of the chassis, handset antenna designs making use of the chassis mode usually have high radiation efficiency [87].

1.3.2.2 Electrically Small Antenna Designs Based on the clear physical investigations of electrically small antenna structures, the CM theory has produced plenty of sophisticated antenna designs. These electrically small antennas include the following:

- Radio-frequency identification (RFID) tag antenna in ultra high frequency (UHF) band [137, 138]
- Universal serial bus (USB) dongle antenna for wireless local area network (WLAN) applications [139]
- Helical spherical antennas [140–142]
- Microstrip antennas [143–146]
- Loop antenna [147]

The investigations into the CMs of these antenna structures result in a clear understanding of the natural resonance frequency, the radiating current behavior, the pattern and polarization behavior of the radiating fields in the near- and far-field zone. All these valuable investigations serve as principal guidelines in feeding designs. These CM-based antenna designs feature one or more of the following enhanced performances:

- Improved radiation efficiency and gain
- Enhanced bandwidth or multi-band performance
- Improved polarization purity
- Reduced antenna volume
- Simplified feeding designs

1.3.2.3 Platform Antenna Designs   Communication antennas play a vital role in a variety of moving platforms including aircrafts [148], unmanned aerial vehicles (UAV) [149–154], ships [155], and land vehicles [69, 70, 156]. Because of the moving property, long distance transmission is generally required. In order to achieve long distance transmission, the antennas have to work at relatively low-frequency bands, such as the HF, VHF, and UHF bands. The long wavelength of the electromagnetic waves results in a large antenna size if conventional antenna design concepts are to be adopted. In consideration of the aerodynamic property and the aesthetic requirement, antennas with low profile, compact sizes or even be conformal with the platform are in demand, although this is a challenging task. The CM theory contributes a novel way to address the challenges in such kind of platform-integrated antenna design problems. The basic idea in the CM approach is to exploit the CMs of the platform and invent some miniature elements to excite one or more CMs for designated radiation pattern or wide bandwidth. The CM-based platform antennas feature with low profile or even platform conformal property.

For example, an antenna conformal to the V-shaped tail of the UAV is designed using the CMs of the UAV [152]. It radiates a vertically polarized omnidirectional pattern in the frequency range from 50 to 90 MHz. The vertical dimension of the tail is only $\lambda/17$ at 50 MHz. Evidently, if the CMs of the UAV body was not excited for radiation purpose, it is impossible for an antenna with a size of only $\lambda/17$ to radiate energy to the far-field zone. Moreover, the CM analysis of the UAV shows the optimal location for maximum energy coupling from the feeding structures.

Owing to the diversity of the CMs of platforms, the CM theory provides many design freedoms to achieve various design objectives. In the following, we list some typical CM-based platform antenna design cases:

- UAV antennas with reconfigurable radiation patterns [149, 150]
- Two-port MIMO antennas on UAV [154]
- Shipboard antenna with directional radiation pattern [155]
- Land vehicle–mounted antenna with directional radiation pattern [69, 70, 156]

It should be noted that all of these platform antennas fully make use of the CMs of the platform. As compared to traditional antenna design concept, they possess enhanced radiation efficiency and low-profile property, if conformal is not impossible.

1.3.2.4 Computation of Quality Factors for Antennas   The antenna quality factor Q has been extensively studied for a long time. It is useful to measure the maximum possible antenna bandwidth and radiation efficiency, especially in the analysis of electrically small antennas. There are many approaches to formulate the Q factor of an antenna [12, 13, 157–160]. Recently, the CM theory was used to expand the total Q in terms of the modal quality factor Q_n [161–167]. Two approaches are introduced to compute Q_n from the CMs of an antenna:

- Following the Q factor formulation based on the input admittance [168], Elghannai and Rojas proposed to expand the total input admittance of an antenna in terms of the individual admittance of each CM [161]. The modal quality factor

Q_n was then computed in the form based on the Q formulation in Ref. [168]. The total Q factor can then be obtained as a summation of the modal quality factor Q_n.

- The second approach to compute the Q_n factor was based on the energy and power associated with each CM [162]. The calculation of the modal energies and the Q_n factor allows studying the effect of the radiating shape independent of the feedings. The total Q can be further formulated as a superposition of each modal quality factor Q_n.

Both the approaches are accurate for the total Q computation. Moreover, the modal quality factor, Q_n, gives more physical insights into the radiation capability of each mode. The modal quality factor, Q_n, has found wide applications in the computations of bandwidth and radiation efficiency for mobile handset antennas [163], MIMO antennas [164], Franklin antenna [165], dipole antenna [166], and microstrip antennas [167].

1.3.2.5 Antenna Designs Using Reactive Loading

In most cases, the CMs of an unloaded antenna structure may not meet the expected radiation performance. Therefore, the radiating structure has to be modified to ensure the dominant CMs hold the desired radiation performance in terms of the radiation pattern, current distribution, bandwidth, and so on. Reactive loading is one of the most popular approaches to modify the CMs of an existing radiating structure. The current distribution is modified by adding lump elements such as capacitors and inductors, while the geometry of the original radiating structure does not need further modification.

There are many typical antenna designs produced by reactively loading the antennas according to the characteristic currents over the existing antenna structures:

- Pattern reconfigurable antenna array designs: In Ref. [169], pattern reconfigurable circular arrays with low sidelobe levels are synthesized by reactively loading the arrays. In this approach, the CM analysis for the antenna array was performed first to identify its different radiating modes. Based on these CMs, the differential evolution optimizer [170] was then employed to seek the reactive loading values according to the specific requirements on the radiation patterns. The benefits brought by the CM theory were observed in terms of the quality of the optimal design, the convergence rate, and the computational complexity.
- Yagi-Uda antenna designs: The CM theory was combined with the differential evolution optimizer for the optimal design of reactively loaded Yagi-Uda antenna [171]. Based on the CM analysis, the differential evolution algorithm was applied to find the optimal values of the reactance loads for the individual elements in a Yagi-Uda antenna. With the reactive loading, optimal Yagi-Uda antenna designs in terms of the antenna gain, input impedance, and sidelobe level were realized.
- Broadband antenna designs: In Refs. [172–174], a CM-based approach was proposed to systematically design antennas with broadband impedance and pattern characteristics using reactive loadings. Antennas of arbitrary geometry

have their own bandwidths. It was shown that the ideally desired current distribution of an antenna over a wide frequency range could be achieved by using a finite number of loadings. Enhanced bandwidth of a typical narrow band dipole antenna was observed through using this CM-based approach.

• Frequency reconfigurable antenna designs: As a further extension to the work in Ref. [172], the CM theory was applied in frequency reconfigurable antenna designs [175]. It was demonstrated that the reactive loadings could be determined systematically to resonate the antenna at many frequency points over a wide frequency range. A dipole antenna with a wide tuning range of 1:4 bandwidth is achieved by simply tuning it at four loading ports. This approach was also implemented in a frequency reconfigurable planar inverted-F antenna design.

1.3.2.6 Antenna Shape Synthesis The performance of an antenna is dependent on the antenna geometry. At the University of Ottawa, Canada, research on the antenna shape synthesis has been going for many years [176–178]. Owing to the source independent solution of the CM analysis, the proposed antenna shape synthesis allows shaping of the antenna geometry prior to any specific feeding structures. It reduces the constraints placed on the optimization process and leads new designs. The CM-based shape synthesis approach intrinsically ensures the optimized structures having efficient radiating modes for best impedance matching at a given frequency. At the end of the antenna shape optimization, an optimal feed point can be easily determined from the modal currents. Reported examples demonstrate that the quality factor Q of the resulting shaped antenna closely approaches the fundamental Q bounds.

The CM research group at the University of Ottawa has also developed a sub-structure CM concept [179]. It extended the applicability of the antenna shape-synthesis technique where only parts of the entire antenna structure need shape synthesis.

1.3.2.7 Planar Monopole Antennas With regard to many of its attractive features, like wide frequency band, omnidirectional pattern, compact size, and low cost, the planar monopole antennas are very popular in many wireless communication systems and UWB systems. The CM analysis provides interesting physical insights into the radiation phenomena taking place in the planar monopole antennas. It revealed that the wideband performance could be characterized by the modal voltage-standing-wave ratio, modal current distribution, and modal significance [180]. The modal analysis also showed that the dominant mode controls the antenna's behavior in the lower band, and the higher order modes control the behavior in the upper band [181–183].

Modal analysis of band-notched planar monopoles illustrates that the resonant modes are due to the embedded narrowband slot structure [184]. By electronically controlling the excitation of the slot mode, planar monopoles with switchable band-notched behavior can be obtained. Similarly, a tunable band-notched UWB antenna can be realized by controlling the resonance of this slot mode. As mode excitation is proportional to current amplitude at the feed point, the bandwidth can be improved by properly combining and exciting more than one mode through multiple excitation sources. These in-depth understandings pave the way for the proposal of novel designs of planar monopole antennas through the control of the excitation and resonance of specific modes [185].

1.3.2.8 CM for Structures with Dielectric Materials   CM analysis for structures with dielectric materials is far more complicated than PEC problems. Specifically, the volume integral equation raises large number MoM unknowns and the CM analysis is time-consuming, even for an electrically small dielectric body [186]. As the surface integral equation is invoked for CM analysis, it involves the surface electric currents and magnetic currents. These two types of currents are dependent on each other. This may lead to some unphysical modes if the MoM matrix is directly used for the CM analysis of structures with dielectric materials [31, 33, 187, 188]. The CM analysis of structures with dielectric materials will be further discussed in Chapter 4.

1.3.2.9 Equivalent Circuit Model of Antennas   Physics based on equivalent circuit models are helpful in the design and optimization of antennas. As opposed to the mathematically fitted model, the equivalent circuit model derived from the CM theory provides physical insights and allows the exploration of the limits of attainable performance. The representative research on the developments of equivalent circuit model for antennas using the CM theory includes:

- Equivalent circuit model for conducting chassis with capacitive and inductive coupling elements was developed [189]. In this model, most of the elements can be directly obtained without numerical optimization. The equivalent circuit model developed found its applications in the design of mobile handset antennas.
- The series and parallel resonances of an antenna's input impedance are explained from the point view of CMs [190]. It showed that the parallel resonance corresponds to the interaction of at least two nearby CMs. The eigenvalues of the two modes have opposite signs. On the other hand, the series resonance is mainly contributed by the resonance of a single CM. This finding is supported by the CM analysis for a simple wire dipole antenna, an edge-fed patch antenna, and a loop antenna.
- An approach for modeling antenna impedances and radiation fields using fundamental eigenmodes was developed in Refs. [191, 192]. Higher order modes can be more accurately modeled with added circuit complexity. Owing to the physical behavior of the CMs, the developed model accurately links the circuit models with radiation patterns and other field behavior, and the far-field patterns and antenna gain of a dipole could be accurately extrapolated over a 10:1 bandwidth.
- Very recently, Adams showed that the frequency response of each CM could be approximated by a template function related to the spherical mode [193]. By choosing the appropriate template function, the frequency response of the CMs can be modeled over a broadband. Based on this observation, the acceleration in the interpolation of antenna's impedance could be achieved over a wide bandwidth.

1.3.2.10   Radiation Pattern Synthesis and Decomposition   CMs consist of a complete set of orthogonal modes. The far-fields of each mode are orthogonal with each other. Any radiating far-fields can be expressed as a superposition of the CMs. This property is known as the modal solution of CMs. The synthesis and decomposition of radiation patterns is a direct application of the modal solution and orthogonal property of the CMs.

In radiation pattern synthesis problem, the objective is to find the optimal weightings for each of the CMs, such that these weightings results in a radiation pattern with satisfactory specifications. Recently, a CM-based radiation pattern synthesis procedure was developed by using a multiobjective optimizer [149, 194]. The weightings for each of the CMs also provided the corresponding radiating currents.

In radiation pattern decomposition problem, however, the objective is to express a given radiation pattern into a set of CMs. In Refs. [195, 196], this far-field decomposition problem was also called reconstruction of the CMs from radiated far fields. The far-field decomposition showed that the radiation mechanism of a complicated antenna structure could be approximated by a simplified structure. The weighting coefficients of all significant modes could be reconstructed with good accuracy even for complex real structures such as mobile phones [195]. As a consequence, the far-field decomposition provides a new way to understand the radiation mechanism of antennas with very complicated geometries.

1.3.2.11   Modal Tracking Algorithms   Practical wideband CM analysis requires the modes at one frequency to be associated with the modes at another frequency. The association relationships of eigenvectors at two different frequencies are generally determined from the correlation coefficients among the eigenvectors. The concept of modal tracking is introduced in the CM theory to sort the modes in the correct order at each frequency. Many tracking algorithms were developed to address this problem [197–200]. The sorted CMs clearly showed the evolution of the modes from non-resonant frequency to resonant frequency and vice versa. These modal tracking algorithms provide a new perspective to investigate the resonant behaviors of CMs. The details of the modal tracking algorithms will be discussed in Chapter 2.

1.4   CHARACTERISTIC MODES IN SCATTERING COMPUTATION

In addition to the wide applications of CM theory in antenna engineering, the CM theory is also applicable to scattering computations, especially in the multiple scattering problems. For simple targets, in most cases, a few low-order CMs are sufficient to reproduce a good approximation to the induced current on scatterers. For a multiple scattering problem consists of large number of small objects, the CMs on each object can be used as the entire domain basis function. Therefore, the MoM using these entire domain CM basis functions involves a significantly reduced number of unknowns. For this reason, it allows the analysis of large multiple scattering problems using direct methods.

At Università della Calabria, Italy, Massa and his colleagues carried out many interesting works to solve multiple scattering problems using CMs [201–211]. Their research began with the multiple scattering by arbitrary-shaped conducting cylinders [201–203]. The use of CMs as entire domain basis functions in the MoM was proved to be very effective in the analysis of the scattering from collections of arbitrarily shaped cylinders. As an extension to their previous study, CMs of elliptic cylinder were used in the computation of the TM scattering from a collection of elliptic cylinders [204]. Both computational efficiency and accuracy were demonstrated through the multiple scattering computations.

Following the same idea, the scattering from large microstrip antenna arrays was simulated by using the CMs of each antenna element as entire domain basis functions. In addition to the scattering analysis, mutual coupling [207] and radiation performance [208] of large microstrip antenna arrays were also addressed by using the CMs as entire domain basis functions in the MoM.

1.5 OUTLINE OF THIS BOOK

With the great support from Temasek Laboratories at National University of Singapore, the authors have devoted great effort in the theoretical developments as well as CM-based methodology development for a variety of critical antenna designs. Our studies on CM theory along with those from other research groups are scattered throughout many technical reports, journal, and conference papers. The intention of this book is to compile and organize these advanced research achievements made in the area of CM studies. It is the first comprehensive book on the CM theory and its applications. Thus, it will be an invaluable book for antenna researchers, engineers, and students who are seeking for antenna design methodology and concept with clear underlying physics in their antenna research and development.

Characteristic Modes: Theory and Applications in Antenna Engineering is organized into six chapters. Chapter 1 gives an introduction and review of the history developments as well as the recent advances of the CM theory and its applications in antenna engineering.

In Chapter 2, the CM theory for PEC bodies is first formulated using the EFIE. CM formulations based on the magnetic field integral equation, approximate magnetic field integral equation, and combined field integral equation (CFIE) are then discussed. This chapter also addresses the numerical implementation of the CM theory. Numerical techniques for solving generalized eigenvaule equation and tracking CMs across a wide frequency band are discussed. The physics of the characteristic modes and its relationship with the spherical modes are illustrated through numerical examples. Numerical aspects of CM computation are also discussed. Finally, the CMs of a planar inverted-F antenna are given to provide a first glance at how to apply the CM theory in practical antenna designs.

Chapter 3 discusses the CM theory for structures embedded in multilayered medium. The underlying physics of the CM theory is revealed through the comparison with the cavity modal solutions of a rectangular patch antenna printed on a dielectric substrate. The CM analysis for a triangular microstrip antenna and a concentric ring antenna is further revealed to show how the presented CM theory can be beneficial to circular polarized and pattern reconfigurable multimode microstrip antenna designs. In addition, circularly polarized microstrip antennas with corner cutting, U-slot, and E-shaped patch are described. These examples illustrate how CM-based designs can enhance the axial ratio and cross-polarization performances. It also shows that the presented CM theory allows engineers to determine the axial ratio bandwidth prior to the feeding structure design. Therefore, one can focus their effort on the feeding design once the CM analysis indicates that the antenna structure has the potential to radiate CP waves in a particular frequency band.

Chapter 4 presents the recent advances in the CM theory for dielectric bodies and its promising applications to the design of dielectric resonant antennas. Two generalized eigenvalue equations based on the PMCHWT surface integral equation formulation are introduced to predict the resonant frequencies of DRAs. Brief discussions on the CM formulation for dielectric bodies proposed in the 1970s are provided to show the essences of our newly developed CM formulations. CM analysis results for cylindrical, spherical, rectangular, notched rectangular and triangular DRAs are presented and discussed. These typical examples obtained clearly demonstrate that the presented CM theory is very promising in the understanding and designing of feeding probes for obtaining certain radiating mode. Guidelines for the excitation of a particular mode using either the coaxial probe feeding or aperture coupling feeding are presented to show how to apply the CM theory in practical DRA designs.

Chapter 5 is devoted to the CM theory for N-port networks and its applications in the optimization of antenna arrays. The completeness and orthogonality properties of the CMs are taken into account in the antenna array optimizations. In comparison to those array optimizations involving full-wave analysis, the efficiency of the CM-based optimization method is improved up to three orders. The mutual coupling is also considered in the CM-based method, and thus the accuracy is kept the same as that in the full-wave analysis. Concentric ring antenna arrays for main beam steering and sidelobe suppression are presented. CM-based optimal design of Yagi-Uda antennas for high directivity and low sidelobe is also presented to show the flexibility of the method. Finally, to improve the bandwidth of tightly coupled antenna arrays, the synthesis of termination conditions for edge elements using the present CM theory is presented.

Chapter 6 discusses the design of platform-integrated antenna systems using CMs. Recent advancements in antenna system designs on aircraft and ship are introduced. Using the CMs of the platforms, the current distributions on the platforms can be efficiently synthesized for any desired radiation patterns. The feedings for the excitation of these currents can either have a very low profile or be conformal to the surface of the platforms. This speciality makes the CM-based method very attractive in many practical applications, especially those operating at the HF, VHF, and UHF frequency bands. Details of the design procedure and the experimental results are presented to show how to make use of the CM theory in these designs.

REFERENCES

[1] R. F. Harrington, *Field Computation by Moment Methods*. New York, NY: Macmillan, 1968.

[2] P. P. Silvester and R. L. Ferrari, *Finite Elements for Electrical Engineers* (3rd edition). Cambridge, UK: Cambridge University Press, 1996.

[3] J.-M. Jin, *The Finite Element Method in Electromagnetics* (2nd edition). New York, NY: John Wiley & Sons, Inc., 2002.

[4] K. S. Yee, "Numerical solution of initial boundary value problems involving Maxwell's equations in isotropic media," *IEEE Trans. Antennas Propag.*, vol. 14, no. 3, pp. 302–307, 1966.

[5] K. S. Kunz and R. J. Luebbers, *The Finite Difference Time Domain Method for Electromagnetics*. Boca Raton, FL: CRC Press, 1994.

[6] A. Taflove and S. C. Hagness, *Computational Electrodynamics: The Finite Difference Time Domain Method* (3rd edition). Norwood, MA: Artech House, 2005.

[7] R. L. Haupt and D. H. Werner, *Genetic Algorithms in Electromagnetics*. Hoboken, NJ: John Wiley & Sons, Inc., 2007.

[8] F. A. Alhargan and S. R. Judah, "Tables of normalized cutoff wavenumbers of elliptic cross section resonators," *IEEE Trans. Microw. Theory Tech.*, vol. 42, no. 2, pp. 333–338, Feb. 1994.

[9] T. V. Khai and C. T. Carson, "m=0, n=0 mode and rectangular waveguide slot discontinuity," *Electron. Lett.*, vol. 9, no. 18, pp. 431–432, Sep. 1973.

[10] S. Li and B. S. Wang, "Field expressions and patterns in elliptical waveguide," *IEEE Trans. Microw. Theory Tech.*, vol. 48, no. 5, pp. 864–867, May 2000.

[11] M. Schneider and J. Marquardt, "Fast computation of modified Mathieu functions applied to elliptical waveguide problems," *IEEE Trans. Microw. Theory Tech.*, vol. 47, no. 4, pp. 573–576, Apr. 1999.

[12] L. J. Chu, "Physical limitations of omnidirectional antennas," *J. Appl. Phys.*, vol. 19, no. 12, pp. 1163–1175, Dec. 1948.

[13] R. F. Harrington, "Effect of antenna size on gain, bandwidth and efficiency," *J. Res. Nat. Bur. Stand.-D, Radio Propag.*, vol. 64D, no. 1, pp. 1–12, Jan.–Feb. 1960.

[14] Y. T. Lo, D. Solomon, and W. F. Richards, "Theory and experiment on microstrip antennas," *IEEE Trans. Antennas Propag.*, vol. AP-27, no. 2, pp. 137–145, Mar. 1979.

[15] W. F. Richards, Y. T. Lo, and D. D. Harrison, "An improved theory for microstrip antennas and applications," *IEEE Trans. Antennas Propag.*, vol. AP-29, no. 1, pp. 38–46, Jan. 1981.

[16] R. K. Mongia and A. Ittipiboon, "Theoretical and experimental investigations on rectangular dielectric resonator antennas," *IEEE Trans. Antennas Propag.*, vol. 45, no. 9, pp. 1348–1356, Sep. 1997.

[17] R. K. Mongia, "Theoretical and experimental resonant frequencies of rectangular dielectric resonators," *IEE Proc.-H*, vol. 139, no. 1, pp. 98–104, Feb. 1992.

[18] Y. M. Pan, K. W. Leung, and K. M. Luk, "Design of the millimeter-wave rectangular dielectric resonator antenna using a higher-order mode," *IEEE Trans. Antennas Propag.*, vol. 59, no. 8, pp. 2780–2788, Aug. 2011.

[19] A. Petosa and S. Thirakoune, "Rectangular dielectric resonator antennas with enhanced gain," *IEEE Trans. Antennas Propag.*, vol. 59, no. 4, pp. 1385–1389, Apr. 2011.

[20] K. W. Leung, E. H. Lim, and X. S. Fang, "Dielectric resonator antennas: From the basic to the aesthetic," *Proc. IEEE*, vol. 100, no. 7, pp. 2181–2193, Jul. 2012.

[21] R. J. Garbacz, "Modal expansions for resonance scattering phenomena," *Proc. IEEE*, vol. 53, no. 8, pp. 856–864, Aug. 1965.

[22] R. J. Garbacz and R. H. Turpin, "A generalized expansion for radiated and scattered fields," *IEEE Trans. Antennas Propag.*, vol. AP-19, no. 3, pp. 348–358, May 1971.

[23] R. J. Garbacz and D. M. Pozar, "Antenna shape synthesis using characteristic modes," *IEEE Trans. Antennas Propag.*, vol. AP-30, no. 3, pp. 340–350, May 1982.

[24] R. F. Harrington and J. R. Mautz, "Theory of characteristic modes for conducting bodies," *IEEE Trans. Antennas Propag.*, vol. AP-19, no. 5, pp. 622–628, Sep. 1971.

[25] R. F. Harrington and J. R. Mautz, "Computation of characteristic modes for conducting bodies," *IEEE Trans. Antennas Propag.*, vol. AP-19, no. 5, pp. 629–639, Sep. 1971.

[26] J. R. Mautz and R. F. Harrington, "Modal analysis of loaded N-port scatterers," *IEEE Trans. Antennas Propag.*, vol. AP-21, no. 2, pp. 188–199, Mar. 1973.

[27] A. Krewski and W. L. Schroeder, "N-Port DL-MIMO antenna system realization using systematically designed mode matching and mode decomposition network," in European Microwave Conference 2012, Amsterdam, the Netherlands, Oct. 2012.

[28] I. Tzanidis, K. Sertel, and J. L. Volakis, "Characteristic excitation taper for ultrawideband tightly coupled antenna arrays," *IEEE Trans. Antennas Propag.*, vol. 60, no. 4, pp. 1777–1784, Apr. 2012.

[29] Y. Chang and R. F. Harrington, "A surface formulation for characteristic modes of material bodies," *IEEE Trans. Antennas Propag.*, vol. AP-25, no. 6, pp. 789–795, Nov. 1977.

[30] R. F. Harrington, J. R. Mautz, and Y. Chang, "Characteristic modes for dielectric and magnetic bodies," *IEEE Trans. Antennas Propag.*, vol. AP-20, no. 2, pp. 194–198, Mar. 1972.

[31] Y. Chen and C. F. Wang, "Surface integral equation based characteristic mode formulation for dielectric resonators," in 2014 IEEE International Symposium on Antennas and Propagation, Memphis, TN, pp. 846–847, Jul. 2014.

[32] H. Alroughani, J. Ethier, and D. A. McNamara, "On the classification of characteristic modes, and the extension of sub-structure modes to include penetrable material," in 2014 International Conference on Electromagnetics in Advanced Applications (ICEAA), Aruba, pp. 159–162, Aug. 2014.

[33] H. Alroughani, J. Ethier, and D. A. McNamara, "Observations on computational outcomes for the characteristic modes of dielectric objects," in IEEE AP-S International Symposium Digest, Memphis, TN, pp. 844–845, Jul. 2014.

[34] E. Safin and D. Manteuffel, "Resonance behaviour of characteristic modes due to the presence of dielectric objects," in 7th European Conference on Antennas and Propagation (EUCAP), Gothenburg, Sweden, Apr. 2013.

[35] R. T. Maximidis, C. L. Zekios, T. N. Kaifas, E. E. Vafiadis, and G. A. Kyriacou, "Characteristic mode analysis of composite metal-dielectric structure, based on surface integral equation/moment method," in 8th European Conference on Antennas and Propagation (EUCAP), Hague, the Netherlands, Apr. 2014.

[36] N. Inagaki and R. J. Garbacz, "Eigenfunctions of composite hermitian operators with application to discrete and continuous radiating systems," *IEEE Trans. Antennas Propag.*, vol. AP-30, no. 4, pp. 571–575, Jul. 1982.

[37] D. Liu, R. J. Garbacz, and D. M. Pozar, "Antenna synthesis and optimization using generalized characteristic modes," *IEEE Trans. Antennas Propag.*, vol. 38, no. 6, pp. 862–868, Jun. 1990.

[38] C. M. Butler, Y. Rahmat-Samii, and R. Mittra, "Electromagnetic penetration through apertures in conducting surfaces," *IEEE Trans. Antennas Propag.*, vol. AP-26, no. 1, pp. 82–93, Jan. 1978.

[39] H. A. Bethe, "Theory of diffraction by small holes," *Phys. Rev.*, vol. 66, no. 7–8, pp. 163–182, Oct. 1944.

[40] K. Y. Kabalan, R. F. Harrington, J. R. Mautz, and H. A. Auda, "Characteristic modes for slots in a conducting plane, TE case," *IEEE Trans. Antennas Propag.*, vol. AP-35, no. 2, pp. 162–168, Feb. 1987.

[41] K. Y. Kabalan, R. F. Harrington, J. R. Mautz, and H. A. Auda, "Characteristic modes for slots in a conducting plane, TM case," *IEEE Trans. Antennas Propag.*, vol. AP-35, no. 3, pp. 331–335, Mar. 1987.

[42] K. Y. Kabalan, A. El-Hajj, and R. F. Harrington, "Characteristic mode analysis of a slot in a conducting plane separating different media," *IEEE Trans. Antennas Propag.*, vol. AP-38, no. 4, pp. 476–481, Apr. 1990.

[43] A. El-Hajj, K. Y. Kabalan, and R. F. Harrington, "Characteristic modes of a slot in a conducting cylinder and their use for penetration and scattering, TE case," *IEEE Trans. Antennas Propag.*, vol. AP-40, no. 2, pp. 156–161, Feb. 1992.

[44] K. Y. Kabalan, A. El-Hajj, and R. F. Harrington, "Characteristic modes of a slot in a conducting cylinder and their use for penetration and scattering, TM case," *IEE Proc.-H, Microw. Antennas Propag.*, vol. 139, no. 3, pp. 287–291, Jun. 1992.

[45] K. Y. Kabalan, A. El-Hajj, and R. F. Harrington, "Scattering and penetration characteristic mode for a dielectric filled conducting cylinder with longitudinal slot," *AEU J. Electron. Commun.*, vol. 47, no. 3, pp. 137–142, Jun. 1993.

[46] A. El-Hajj, K. Y. Kabalan, and R. F. Harrington, "Characteristic modes analysis of electromagnetic coupling through multiple slots in a conducting plane," *IEE Proc.-H Microw. Antennas Propag.*, vol. 40, no. 6, pp. 421–425, Dec. 1993.

[47] K. Y. Kabalan and A. El-Hajj, "Characteristic mode formulation of the aperture-fed waveguide problem," *AEU J. Electron. Commun.*, vol. 48, no. 2, pp. 130–134, Feb. 1994.

[48] A. El-Hajj and K. Y. Kabalan, "Scattering from and penetration into a dielectric conducting cylinder with multiple apertures," *IEEE Trans. Electromagn. Compat.*, vol. 36, no. 3, pp. 196–200, Aug. 1994.

[49] A. El-Hajj and K. Y. Kabalan, "Characteristic modes of a rectangular aperture in a perfectly conducting plane," *IEEE Trans. Antennas Propaga.*, vol. 42, no. 10, pp. 1447–1450, Oct. 1994.

[50] A. El-Hajj, K. Y. Kabalan, and S. V. Khoury, "Electromagnetic coupling between two half-space regions separated by multiple slot-perforated parallel conducting screens," *IEEE Trans. Electromagn. Compat.*, vol. 37, no. 1, pp. 105–109, Feb. 1995.

[51] K. Y. Kabalan and A. El-Hajj, "A CM solution of the parallel plate-fed slot antenna," *Radio Sci.*, vol. 30, no. 2, pp. 353–360, Mar./Apr. 1995.

[52] A. E-Hajj, K. Y. Kabalan, and A. Rayes, "Three-dimensional characteristic mode formulation of the cavity-backed aperture problem," *AEU Int. J. Electron. Commun.*, vol. 50, no. 3, pp. 208–214, May 1996.

[53] K. Y. Kabalan, A. El-Hajj, S. Khoury, and A. Rayes, "Electromagnetic coupling to conducting objects behind apertures in a conducting body," *Radio Sci.*, vol. 32, no. 3, pp. 881–898, May/Jun. 1997.

[54] S. Khoury, K. Y. Kabalan, A. El-Hajj, and A. Rayes, "Electromagnetic radiation and scattering at the american university of beirut, feature article," *IEEE Antennas and Propag. Mag.*, vol. 39, no. 3, pp. 40–43, Jun. 1997.

[55] A. El-Hajj, K. Y. Kabalan, and A. Rayes, "Characteristic mode formulation of multiple rectangular aperture in a conducting plane with a dielectric filled cavity," *IEEE Trans. Electromagn. Compati.*, vol. 40, no. 2, pp. 89–93, May 1998.

[56] A. El-Hajj, K. Y. Kabalan, and A. Rayes, "Electromagnetic scattering and penetration into a filled cavity through an aperture in a conducting plane," *Radio Sci.*, vol. 33, no. 5, pp. 1267–1275, Sep./Oct. 1998.

[57] A. El-Hajj, K. Y. Kabalan, and A. Rayes, "Electromagnetic transmission between two regions separated by two parallel planes perforated with rectangular apertures," *AEU J. Electron. Commun.*, vol. 54, no. 4, pp. 203–209, Aug. 2000.

[58] K. Y. Kabalan, A. El-Hajj, and A. Rayes, "A three dimensional characteristic mode solution of two perforated parallel planes separating different dielectric mediums," *Radio Sci.*, vol. 36, no. 2, pp. 183–194, Mar.–Apr. 2001.

[59] K. Y. Kabalan, A. El-Hajj, and A. Rayes, "Electromagnetic penetration through three different dielectric regions separated by two parallel planes perforated with multiple apertures," *Iran. J. Electr. Comput. Eng.*, vol. 2, no. 1–2, pp. 69–74, Jun. 2003.

[60] K. Y. Kabalan, A. El-Hajj, and R. F. Harrington, "Complete study of electromagnetic coupling between two regions through narrow slot," in Proceedings of the 1989 URSI International Symposium on EM Theory, Stockholm, Sweden, pp. 422–424, Aug. 1989.

[61] K. Y. Kabalan, A. El-Hajj, and A. Rayess, "A generalized network formulation for a circular aperture in a conducting plane," in Proceedings of the Nineteenth National radio Science Conference, Alexandria, Egypt, pp. 81–89, Mar. 9–21, 2002.

[62] A. El-Hajj, K. Y. Kabalan, and F. Elias, "The characteristic mode theory of the field penetration into a cylindrical waveguide through a rectangular aperture, TE case," in Proceedings of the 1st International Conference on Information & Communication Technologies: from Theory to Applications—ICTTA'04, Damascus, Syria, pp. 239–240, Apr. 19–23, 2004.

[63] K. Y. Kabalan, A. El-Hajj, and F. Elias, "The characteristic mode theory of the field penetration into a cylindrical waveguide through a rectangular aperture, TM case," in Proceedings of the 2004 International Conference on Electrical, Electronic, and Computer Engineering, ICEEC'04, Cairo, Egypt pp. 585–588, Sep. 5–7, 2004.

[64] B. V. Elsevier, Homepage of Scoupus database. Available at http://www.scopus.com. Accessed January 24, 2015.

[65] E. H. Newman, "Small antenna location synthesis using characteristic modes," *IEEE Trans. Antennas Propag.*, vol. AP-25, no. 4, pp. 530–795, Jul. 1977.

[66] R. F. Harrington and J. R Mautz, "Control of radar scattering by reactive loading," *IEEE Trans. Antennas Propag.*, vol. AP-20, no. 4, pp. 446–454, Jul. 1972.

[67] A. Yee and R. J. Garbacz, "Self- and mutual-admittances of wire antennas in terms of characteristic modes," *IEEE Trans. Antennas Propag.*, vol. 21, no. 6, pp. 868–871, Nov. 1973.

[68] R. F. Harrington and J. R. Mautz, "Optimization of radar cross section of N-port loaded scatterers," *IEEE Trans. Antennas Propag.*, vol. AP-22, no. 5, pp. 697–701, Sep. 1974.

[69] B. A. Austin and K. P. Murray, "The application of characteristic-mode techniques to vehicle-mounted NVIS antennas," *IEEE Antennas Propag. Mag.*, vol. 40, no. 1, pp. 7–21, Feb. 1998.

[70] K. P. Murray and B. A. Austin, "Synthesis of vehicular antenna NVIS radiation patterns using the method of characteristic modes," *IEE Proc. Microw. Antennas Propag.*, vol. 141, no. 3, pp. 151–154, Jun. 1994.

[71] R. F. Harrington and J. R. Mautz, "Characteristic modes for aperture problems," *IEEE Trans. Microw. Theory Tech.*, vol. 33, no. 6, pp. 500–505, Jun. 1985.

[72] Z. Miers, H. Li, and B. K. Lau, "Design of bezel antennas for multiband MIMO terminals using characteristic modes," in Proceedings of the 8th European Conference Antennas and Propagation (EuCAP'2014), Hague, the Netherlands, Apr. 6–10, 2014.

[73] H. Li, Z. Miers, and B. K. Lau, "Generating multiple characteristic modes below 1 GHz in small terminals for MIMO antenna design," in Proceedings of the IEEE International Symposium Antennas and Propagation (APS'2013), Orlando, FL, Jul. 7–13, 2013.

[74] Z. Miers, H. Li, and B. K. Lau, "Design of multi-antenna feeding for MIMO terminals based on characteristic modes," in Proceedings of the IEEE International Symposium Antennas and Propagation (APS'2013), Orlando, FL, Jul. 7–13, 2013.

[75] H. Li, Z. Miers, and B. K. Lau, "Design of orthogonal MIMO handset antennas based on characteristic mode manipulation at frequency bands below 1 GHz," *IEEE Trans. Antennas Propag.*, vol. 62, no. 5, pp. 2756–2766, May 2014.

[76] Z. Miers, H. Li, and B. K. Lau, "Design of bandwidth enhanced and multiband MIMO antennas using characteristic modes," *IEEE Antennas Wirel. Propag. Lett.*, vol. 12, pp. 1696–1699, 2013.

[77] Z. Miers and B. K. Lau, "Design of multimode multiband antennas for MIMO terminals using characteristic mode analysis," in Proceedings of the IEEE International Symposium Antennas and Propagation (APS'2014), Memphis, TN, pp. 1429–1430, Jul. 6–12, 2014.

[78] R. Martens and D. Manteuffel, "A feed network for the selective excitation of specific characteristic modes on small terminals," in 2012 6th European Conference on Antennas and Propagation (EUCAP), Prague, Czech Republic, pp. 1842–1846, Mar. 2012.

[79] R. Martens and D. Manteuffel, "2-Port antenna based on the selective excitation of characteristic modes," in APS 2012—IEEE International Symposium on Antennas and Propagation, Chicago, IL, pp. 1–2, Jul. 2012.

[80] R. Martens and D. Manteuffel, "3-Port MIMO antenna based on the selective excitation of characteristic modes on small terminals," in COST IC1102, WG Meeting & Technical Workshop 2012, Istanbul, Turkey, Sep. 2012.

[81] R. Martens, E. Safin, and D. Manteuffel, "Selective excitation of characteristic modes on small terminals," in Proceedings of the 5th European Conference on Antennas and Propagation (EUCAP), Rome, Italy, pp. 2492–2496, Apr. 2011.

[82] R. Martens, E. Safin, and D. Manteuffel, "Inductive and capacitive excitation of the characteristic modes of small terminals," in Loughborough Antennas and Propagation Conference (LAPC), 2011 Loughborough, UK, pp. 1–4, Nov. 2011.

[83] K. K. Kishor and S. V. Hum, "Multi-feed chassis-mode antenna with dual-band mimo operation," in Proceedings IEEE International Symposium Antennas Propagation (APS'2014), Memphis, TN, pp. 1427–1428, Jul. 2014.

[84] K. K. Kishor and S. V. Hum, "A reconfigurable chassis-mode MIMO Antenna," in 2013 Eur. Conf. Ant. Propag. (EuCAP 2013), Gothenburg, Sweden, pp. 1992–1996, Apr. 2013.

[85] K. K. Kishor and S. V. Hum, "A pattern reconfigurable chassis-mode MIMO antenna," *IEEE Trans. Antennas Propag.*, vol. 62, no. 6, pp. 3290–3298, Jun. 2014.

[86] P. Miskovsky and A. von Arbin, "Evaluation of MIMO handset antennas with decorative metal elements using characteristic modes," in 2014 IEEE AP-Symposium, Memphis, TN, pp. 1423–1424, Jul. 2014.

[87] J. Rahola and J. Ollikainen, "Optimal antenna placement for mobile terminals using characteristic mode analysis," in Proceedings of the EuCAP 2006 Conference, Nice, France, Nov. 6–10, 2006.

[88] R. Martens, and D. Manteuffel, "Systematic design method of a mobile multiple antenna system using the theory of characteristic modes," *IET Microw. Antennas Propag.*, vol. 8, no. 12, pp. 887–893, Sep. 2014.

[89] R. Martens and D. Manteuffel, "Mobile LTE-A handset antenna using a hybrid coupling element," in 2014 IEEE AP-Symposium, Memphis, TN, pp. 1419–1420, Jul. 2014.

[90] T. Hadamik, R. Martens, and D. Manteuffel, "Comparison of two matching networks for a LTE2600 handset antenna," in 2013 Loughborough Antennas and Propagation Conference, Loughborough University, Loughborough, UK, pp. 145–148, Nov. 2013.

[91] R. Martens, J. Holopainen, E. Safin, and D. Manteuffel, "Small terminal multi-antenna system based on two different types of non-resonant coupling elements," in COST IC1102, WG Meeting & Technical Workshop 2013, Ghent, Belgium, Sep. 2013.

[92] R. Martens, J. Holopainen, E. Safin, J. Ilvonen, and D. Manteuffel, "Optimal dual-antenna design in a small terminal multi-antenna system," *IEEE Antennas and Wirel. Propag. Lett.*, vol. 12, pp. 1700–1703, Dec. 2013.

[93] D. Manteuffel, "Small terminal antenna concepts for reconfigurable MIMO handsets," in APCAP 2012-Asia Pacific Conference on Antennas and Propagation, Singapore, pp. 15–16, Aug. 2012.

[94] D. Manteuffel and R. Martens, "Multiple antenna integration in small terminals," in 2012 International Symposium on Antennas and Propagation (ISAP 2012), Nagoys, pp. 211–224, Oct./Nov. 2012.

[95] D. Manteuffel, "Small terminal antenna concepts for reconfigurable MIMO handsets," in 2012 IEEE Asia-Pacific Conference on Antennas and Propagation (APCAP), Singapore, pp. 15–16, Aug. 27–29, 2012.

[96] D. Manteuffel and R. Martens, "A concept for MIMO antennas on small terminals based on characteristic modes," in 2011 International Workshop on Antenna Technology (iWAT), Hong Kong, pp. 17–20, Mar. 2011.

[97] R. Martens and D. Manteuffel, "Element correlation of MIMO antennas on small terminals," in 2010 Proceedings of the Fourth European Conference on Antennas and Propagation (EuCAP), pp. 1–5, Apr. 2010.

[98] R. Martens, E. Safin, and D. Manteuffel, "On the relation between the element correlation of antennas on small terminals and the characteristic modes of the chassis," in Loughborough Antennas and Propagation Conference (LAPC), 2010, Loughborough, UK, pp. 457–460, Nov. 2010.

[99] H. Li, Y. Tan, B. K. Lau, Z. Ying, and S. He, "Characteristic mode based tradeoff analysis of antenna-chassis interactions for multiple antenna terminals," *IEEE Trans. Antennas Propag.*, vol. 60, no. 2, pp. 490–502, Feb. 2012.

[100] H. Li, B. K. Lau, Z. Ying, and S. He, "Decoupling of multiple antennas in terminals
       with chassis excitation using polarization diversity, angle diversity and current control,"
       IEEE Trans. Antennas Propag., vol. 60, no. 12, pp. 5947–5957, Dec. 2012.

[101] H. Li, B. K. Lau, and S. He, "Angle and polarization diversity in compact dual-antenna
       terminals with chassis excitation," in Proceedings of the URSI General Assembly
       (URSI GA'2011), Istanbul, Turkey, Aug. 13–20, 2011.

[102] H. Li, B. K. Lau, Y. Tan, S. He, and Z. Ying, "Impact of current localization on the
       performance of compact MIMO antennas," in Proceedings of the 5th European Conference,
       Antennas Propagation (EuCAP'2011), Rome, Italy, pp. 2423–2426, Apr. 11–15, 2011.

[103] H. Li, B. K. Lau, and Z. Ying, "Optimal multiple antenna design for compact MIMO
       terminals with ground plane excitation," in Proceedings of the International Workshop
       Antenna Technology (IWAT '2011), Hong Kong, P. R. China, Mar. 7–9, 2011.

[104] B. K. Lau and Z. Ying, "Antenna design challenges and solutions for compact MIMO
       terminals," in Proceedings of the International Workshop Antenna Technology (IWAT
       '2011), Hong Kong, P. R. China, Mar. 7–9, 2011.

[105] J. Ethier and D. A. McNamara, "The use of generalized characteristic modes in the design
       of MIMO antennas," *IEEE Trans. Magn.*, vol.45, no.3, pp. 1124–1127, Mar. 2009.

[106] J. Ethier and D. A. McNamara, "An interpretation of mode-decoupled MIMO antennas in
       terms of characteristic port modes," *IEEE Trans. Magn.*, vol. 45, no. 3, pp. 1128–1131,
       Mar. 2009.

[107] J. Ethier, E. Lanoue, and D. A. McNamara, "MIMO handheld antenna design
       approach using characteristic mode concepts," *Microw. Opt. Technol. Lett.*, vol. 50,
       no. 7, pp. 1724–1727, Jul. 2008.

[108] B. D. Raines and R. G. Rojas, "Design of multiband reconfigurable antennas," in 4th
       European Conference on Antennas and Propagation (EUCAP), Barcelona, Spain,
       pp. 1–4, Apr. 12–16, 2010.

[109] E. Antonino-Daviu, M. Cabedo-Fabres, M. Ferrando-Bataller, and J. I. Herranz-
       Herruzo, "Analysis of the coupled chassis-antenna modes in mobile handsets," in 2004
       IEEE Antennas and Propagation Society Symposium, Monterey, vol. 3, pp. 2751–2754,
       Jul. 2014.

[110] E. Antonino-Daviu, C. A. Suarez-Fajardo , M. Cabedo-Fabrés, and M. Ferrando-Bataller,
       "Wideband antenna for mobile terminals based on the handset PCB resonance," *Microw.
       Opt. Technol. Lett.*, vol. 48, no. 7, pp. 1408–1411, Apr. 2006.

[111] M. Cabedo-Fabrés, E. Antonino-Daviu, A. Valero-Nogueira, and M. Ferrando-Bataller,
       "Notched radiating ground plane analyzed from a modal perspective," *Frequenz*, vol. 61,
       no. 3–4, pp. 66–70, Mar.–Apr. 2007.

[112] M. Sonkki, M. Cabedo-Fabres, E. Antonino-Daviu, M. Ferrando-Bataller, and E. T.
       Salonen, "Creation of a magnetic boundary condition in a radiating ground plane to
       excite antenna modes," *IEEE Trans. Antennas Propag.*, vol. 59, no. 10, pp. 3579–3587,
       Oct. 2011.

[113] M. Martinez-Vazquez, E. Antonino-Daviu, and M. Cabedo-Fabres, "Miniaturized
       integrated multiband antennas," in Multiband integrated antennas for 4G terminals.
       Boston, MA: Artech House Publishers, Jul. 2008.

[114] E. Antonino-Daviu, M. Cabedo-Fabres, M. Ferrando-Bataller, and M. Gallo; "Design
       of a multimode mimo antenna using the theory of characteristic modes," *Radio Eng.*,
       vol. 18, no. 4, Part 1, pp. 425–430. Dec. 2009.

[115] E. Antonino-Daviu, M. Cabedo-Fabres, B. Bernardo-Clemente, and M. Ferrando-Bataller, "Printed multimode antenna for MIMO systems," *J. Electromagn. Waves Appl.*, vol. 25, no. 14–15, pp. 2022–2032. 2011.

[116] M. Bouezzeddine and W. L. Schroeder, "Parametric study on capacitive and inductive couplers for exciting characteristic modes on CPE," in European Conference on Antennas and Propagation (EuCAP 2014), Hague, the Netherlands, Apr. 2014.

[117] M. Bouezzeddine and W. L. Schroeder, "On the design of wideband matching and decoupling networks for N-port antennas," in European Conference on Antennas and Propagation (EuCAP 2014), Hague, the Netherlands, Apr. 2014.

[118] A. Krewski and W. L. Schroeder, "Electrically switchable multi-port matching network for 2-port MIMO antennas," in European Conference on Antennas and Propagation (EuCAP 2014), Hague, the Netherlands, Apr. 2014.

[119] A. Krewski, W. L. Schroeder, and K. Solbach, "2-port DL-MIMO antenna design for LTE-enabled USB dongles," *IEEE Antennas Wirel. Propag. Lett.*, vol. 12, pp. 1436–1439, 2013.

[120] A. Krewski, W. L. Schroeder, and K. Solbach, "Multi-band 2-port MIMO antenna design for LTE-enabled 1300 monocoque laptop," in Proceedings of the IEEE Antennas and Propagation Society International Symposium Digest, Orlando, FL, paper 1036, Jul. 2013.

[121] M. Bouezzeddine, A. Krewski, and W. L. Schroeder, "A concept study on MIMO antenna systems for small size CPE operating in the TV white spaces," in COST IC1004, 7th MCM, Ilmenau, Germany, TD(13)07051, May 2013.

[122] A. Krewski, W. L. Schroeder, and K. Solbach, "Multi-band 2-port MIMO LTE antenna design for laptops using characteristic modes," in Loughborough Antennas & Propagation Conference (LAPC 2012), Loughborough, UK, Nov. 2012.

[123] A. Krewski, W. L. Schroeder, and K. Solbach, "MIMO LTE antenna design for laptops based on theory of characteristic modes," in 6th European Conference on Antennas and Propagation (EuCAP 2012), Prague, Czech Republic, Mar. 2012.

[124] A. Krewski, W. L. Schroeder, and K. Solbach, "Matching network synthesis for mobile MIMO antennas based on minimization of the total multi-port return loss," in Loughborough Antennas & Propagation Conference (LAPC 2011), Loughborough, UK, Nov. 2011.

[125] A. Krewski, W. L. Schroeder, and K. Solbach, "Bandwidth limitations and optimum low-band LTE MIMO antenna placement in mobile terminals using modal analysis," in 5th European Conference on Antennas and Propagation (EuCAP 2011), Rome, Italy, Apr. 2011.

[126] A. Krewski and W. L. Schroeder, "Equivalent circuit model for closely coupled symmetrical two-port MIMO antennas in small volume," in Proceedings of the IEEE Antennas and Propagation Society International Symposium Digest, Toronto, CA, paper IF315.6, Jul. 2010.

[127] W. L. Schroeder, A. Acuna Vila, and C. Thome, "Extremely small, wideband mobile phone antennas by inductive chassis mode coupling," in Proceedings of the 36th European Microwave Conference, Manchester, UK, pp. 1702–1705, Sep. 2006.

[128] C. Tamgue-Famdie, W. L. Schroeder, and K. Solbach, "Numerical analysis of characteristic modes on the chassis of mobile phones," in Proceedings of the 1st European Conference on Antennas and Propagation—EuCAP 2006, Nice, France, no. 345145, Nov. 2006.

[129] S. K. Chaudhury, W. L. Schroeder, and H. J. Chaloupka, "MIMO antenna system based on orthogonality of the characteristic modes of a mobile device," in 2nd International ITG Conference on Antennas (INICA 2007), Munich, Germany, pp. 58–62, Mar. 2007.

[130] C. T. Famdie, W. L. Schroeder, and K. Solbach, "Optimal antenna location on mobile phones chassis based on the numerical analysis of characteristic modes," in 10th European Conference onWireless Technology, Munich, Germany, pp. 987–990, Oct. 2007.

[131] S. K. Chaudhury, W. L. Schroeder, and H. J. Chaloupka, "Multiple antenna concept based on characteristic modes of mobile phone chassis," in The Second European Conference on Antennas and Propagation (EuCAP), Edinburgh, UK, Nov. 2007.

[132] W. L. Schroeder, P. Schmitz, and C. Thome, "Miniaturization of mobile phone antennas by utilization of chassis mode resonances," in Proceedings of the German Microwave Conference—GeMiC 2006, Karlsruhe, Germany, no. 7b-3, Mar. 2006.

[133] W. L. Schroeder, C. Tamgue Famdie, and K. Solbach, "Utilisation and tuning of the chassis modes of a handheld terminal for the design of multiband radiation characteristics," in IEE Conference on Wideband and Multi-band Antennas and Arrays, Birmingham, UK, Sep. 2005.

[134] R. T. Maximov, C. L. Zekios, and G. A. Kyriacou, "MIMO antenna design exploiting the characteristic modes eigenanalysis," in 32nd ESA Antenna Workshop on Antennas for Space Applications, Noordwijk, the Netherlands, Oct. 05–08, 2010.

[135] K. K. Kishor and S. V. Hum, "A two-port chassis-mode MIMO antenna," *IEEE Antennas Wirel. Propag. Lett.*, vol. 12, pp. 690–693, May 2013.

[136] A. Abdullah Al-Hadi and R. Tian, "Impact of multiantenna real estate on diversity and mimo performance in mobile terminals," *IEEE Antennas Wirel. Propag. Lett.*, vol. 12, pp. 1712–1715, 2013.

[137] R. Rezaiesarlak and M. Manteghi, "On the application of characteristic modes for the design of chipless RFID tags," in IEEE AP-Symposium, Memphis, TN, pp. 1304–1305, Jul. 2014.

[138] E. A. Elghannai and R. G. Rojas, "Characteristic mode analysis and design of passive UHF-RFID tag antenna," in 2014 IEEE International Conference on RFID, Orlando, FL, Apr. 2014.

[139] E. A. Elghannai and R. G. Rojas, "Design of USB dongle antenna for WLAN applications using theory of characteristic modes," *Electron. Lett.*, vol. 50, no. 4, pp. 249–251, Feb. 2014.

[140] J. J. Adams and J. T. Bernhard, "A modal approach to tuning and bandwidth enhancement of an electrically small antenna," *IEEE Trans. Antennas Propag.*, vol. 59, no. 4, pp. 1085–1092, Apr. 2011.

[141] B. D. Obeidat and R. G. Rojas, "Design and analysis of a helical spherical antenna using the theory of characteristic modes," in IEEE-APS International Symposium, San Diego, CA, pp. 1–4, Jul. 5–11, Jul. 2008.

[142] J. J. Adams and J. T. Bernhard, "Bandwidth enhancement of a small antenna by modal superposition," in Proceedings of the 2009 Antenna Applications Symposium, Monticello, IL, pp. 1–15, Sep. 2009.

[143] Y. Chen and C.-F. Wang, "Characteristic-mode-based improvement of circularly polarized U-slot and E-shaped patch antennas," *IEEE Antennas Wirel. Propag. Lett.*, vol. 11, pp. 1474–1477, 2012.

[144] P. Hazdra, M. Capek, P. Hamouz, and M. Mazanek, "Advanced modal techniques for microstrip patch antenna analysis," in Conference Proceedings ICECom 2010 Zagreb: KoREMA, Dubrovnik, Croatia, pp. 1–6, Sep. 2010.

[145] J. Eichler, P. Hazdra, M. Capek, T. Korinek, and P. Hamouz, "Design of a dual band orthogonally polarized L-probe fed fractal patch antenna using modal methods," *IEEE Antennas Wirel. Propag. Lett.*, vol. 10, p. 1389–1392, 2011.

[146] P. Hazdra, M. Capek, J. Eichler, T. Korinek, and M. Mazanek, "On the modal resonant properties of microstrip antennas," in EuCAP 2012—6rd European Conference on Antennas and Propagation, Prague, Czech Republic, pp. 1650–1654, Mar. 2012.

[147] M. Cabedo-Fabrés, M. Ferrando-Bataller, and A. Valero-Nogueira, "Systematic study of elliptical loop antennas using characteristic modes," in IEEE Antennas and Propagation Society International Symposium, San Antonio, TX, vol. 1, pp. 156–159, Jun. 2002.

[148] J. Chalas, K. Sertel, and J. L. Volakis, "NVIS synthesis for electrically small aircraft using characteristic modes," in 2014 IEEE International Symposium on Antennas and Propagation and USNC-URSI National Radio Science Meeting, Memphis, TN, July 6–12, 2014.

[149] Y. Chen and C.-F. Wang, "Electrically small UAV antenna design using characteristic modes," *IEEE Trans. Antennas Propag.*, vol. 62, no. 2, pp. 535–545, Feb. 2014.

[150] Y. Chen and C.-F. Wang, "From antenna design to feeding design: a review of characteristic modes for radiation problems," in 2013 International Symposium on Antennas and Propagation, Nanjing, China, pp. 68–70, Oct. 2013.

[151] K. Obeidat, R. G. Rojas, and B. Raines, "Design of antenna conformal to V-shaped tail of UAV based on the method of characteristic modes," in 3rd European Conference on Antennas and Propagation (EUCAP), Berlin, Germany, pp. 1–4, March 23–27, 2009.

[152] K. A. Obeidat, B. D. Raines, and R. G. Rojas, "Design of omni-direction vertically polarized vee-shaped antenna using characteristic modes," in 2008 URSI National Meeting, Boulder, CO, Jan. 4–7, 2008.

[153] J. Chalas, K. Sertel, and J. L. Volakis, "Design of in-Situ Antennas Using Platform Characteristic Modes," in 2013 IEEE International Symposium Antennas and Propagation (APS/URSI), Orlando, FL, Jul. 7–13, 2013.

[154] A. Krewski, "2-port MIMO antenna system for high frequency data link with UAVs using characteristic modes," in Proceedings IEEE Antennas and Propagation Society International Symposium Digest, Orlando, FL, Jul. 2013.

[155] Y. Chen and C.-F. Wang, "Shipboard NVIS radiation system design using the theory of characteristic modes," in 2014 IEEE International Symposium on Antennas and Propagation, Memphis, TN, pp. 852–853, Jul. 2014.

[156] M. Ignatenko and D. Filipovic, "Application of characteristic mode analysis to HF low profile vehicular antennas," in IEEE AP-Symposium, Memphis, TN, pp. 850–851, Jul. 2014.

[157] H. A. Wheeler, "Fundamental limitations of small antennas," *Proc. IRE*, vol. 35, pp. 1479–1484, Dec. 1947.

[158] R. E. Collin and S. Rotchild, "Evaluation of antenna Q," *IEEE Trans. Antennas Propag.*, vol. 12, pp. 23–27, Jan. 1964.

[159] J. S. McLean, "A re-examination of the fundamental limits on the radiation Q of electrically small antennas," *IEEE Trans. Antennas Propag.*, vol. 44, pp. 672–675, May 1996.

[160] G. Wen, "A method for the evaluation of small antenna Q," *IEEE Trans. Antennas Propag.*, vol. 51, no. 8, pp. 2124–2129, Aug. 2003.

[161] E. A. Elghannai and R. G. Rojas, "Study of the total Q of antennas in terms of modal qn provided by characteristic modes theory," in 2014 International Symposium on Antennas and Propagation Society, Memphis, TN, pp. 49–50, Jul. 2014.

[162] M. Capek, P. Hazdra, and J. Eichler, "A method for the evaluation of radiation Q based on modal approach," *IEEE Trans. Antennas Propag.*, vol. 60, no. 10, pp. 4556–4567, Oct. 2012.

[163] L. Huang, W. L. Schroeder, and P. Russer, "Estimation of maximum attainable antenna bandwidth in electrically small mobile terminals," in Proceedings of the 36th European Microwave Conference, Manchester, UK, pp. 630–633, Sep. 2006.

[164] E. Safin, R. Martens, M. Capek, J. Eichler, and P. Hazdra, "Discussion and evaluation of modal and radiation Q-factors for MIMO antennas based on theory of characteristic modes," in COST IC1102, 3. WG Meeting & Technical Workshop 2013, Thessaloniki, Greece, Jun. 2013.

[165] P. Hazdra, M. Polivka, M. Capek, and M. Mazanek, "Radiation efficiency and q factor study of franklin antenna using the theory of characteristc modes," in 2011 IEEE Proceedings of the 5th European Conference on Antennas and Propagation, Rome, Italy, pp. 1974–1977, Apr. 2011.

[166] P. Hazdra, M. Capek, J. Eichler, P. Hamouz, and M. Mazanek, "Radiation Q of dipole modal currents," in Proceedings of the 5th European Conference on Antennas and Propagation, Rome, Italy, pp. 1578–1581, Apr. 2011.

[167] J. Eichler, P. Hazdra, M. Capek, and M. Mazanek, "Modal resonant frequencies and radiation quality factors of microstrip antennas," *Int. J. Antennas Propag.*, vol. 2012, 9 p, 2012.

[168] A. D. Yaghjian and S. R. Best, "Impedance, bandwidth and Q of antennas," *IEEE Trans. Antennas Propag.*, vol. 53, no. 4, pp. 1298–1324, Apr. 2005.

[169] Y. Chen and C.-F. Wang, "Synthesis of reactively controlled antenna arrays using characteristic modes and DE algorithm," *IEEE Antennas Wirel. Propag. Lett.*, vol. 11, pp. 385–388, 2012.

[170] Y. Chen, S. Yang, and Z. Nie, "Synthesis of uniform amplitude thinned linear phased arrays using the differential evolution algorithm," *Electromagnetics*, vol. 27, no. 5, pp. 287–297, Jun.–Jul. 2007.

[171] Y. Chen and C.-F. Wang, "Electrically loaded Yagi-Uda antenna optimizations using characteristic modes and differential evolution," *J. Electromagn. Waves Appl.*, vol. 26, pp. 1018–1028, 2012.

[172] K. Obeidat, B. D. Raines, and R. G. Rojas, "Application of characteristic modes and non-foster multiport loading to the design of broadband antennas," *IEEE Trans. Antennas Propag.*, vol. 58, no. 1, pp. 203–207, Jan. 2010.

[173] R. G. Rojas, A. Elfrgani, and E. Elghannai, "Distributed impedance matching with foster and non-foster elements," in 2013 EuCAP, Gothenburg, Sweden, Apr. 2013.

[174] E. A. Elghannai and R. G. Rojas, "Novel multiport non-foster loading technique for wide band antennas," in IEEE USNC/URSI National Radio Science Meeting, Chicago, IL, Jul. 2012.

[175] K. Obeidat, B. D. Raines, R. G. Rojas, and B. T. Strojny, "Design of frequency reconfigurable antennas using the theory of characteristic modes," *IEEE Trans. Antennas Propag.*, vol. 58, no. 10, pp. 3106–3113, Oct. 2010.

[176] J. Ethier and D. A. McNamara, "Antenna shape synthesis without prior specification of the feedpoint," *IEEE Trans. Antennas Propag.*, vol. 62, no. 10, pp. 4919–4934, Oct. 2014.

[177] J. Ethier and D. A. McNamara, "Further applications of a characteristic mode based antenna shape synthesis method," in Proceedings of the 2012 IEEE International Conference on Wireless Information Technology and Systems (ICWITS'2012), Maui, Hawaii, Nov. 2012.

[178] J. Ethier and D. A. McNamara, "Multiband antenna synthesis using characteristic mode indicators as an objective function for optimization," in Proceedings of the 2010 IEEE International Conference on Wireless Information Technology and Systems (ICWITS'2010), USA, Aug./Sep. 2010.

[179] J. Ethier and D. A. McNamara, "A sub-structure characteristic mode concept for antenna shape synthesis," *Electron. Lett.*, vol. 48, no. 9, Apr. 2012.

[180] W. Wu and Y. P. Zhang, "Analysis of ultra-wideband printed planar quasi-monopole antennas using the theory of characteristic modes," *IEEE Antennas Propag. Mag.*, vol. 52, no. 6, pp. 67–77, Dec. 2010.

[181] E. Antonino, M. Cabedo, M. Ferrando, and A. Valero, "Novel wide-band double-fed planar monopole antennas," *Electron. Lett.*, vol. 39, pp.1635–1636, Nov. 2003.

[182] M. Ferrando, M. Cabedo, E. Antonino, and A. Valero, "Overview of planar monopole antennas for uwb applications," in 1st European Conference on Antennas and Propagation, Nice, France, Nov. 2006.

[183] M. Cabedo-Fabres, E. Antonino-Daviu, A. Valero-Nogueira, and M. Ferrando-Bataller, "Analysis of wide band planar monopole antennas using characteristic modes," in 2003 IEEE Antennas and Propagation Society International Symposium, Columbus, OH, vol. 3, pp. 733–736, Jun. 22–27, 2003.

[184] E. Antonino-Daviu, M. Cabedo-Fabres, M. Ferrando-Bataller, and V. M. Rodrigo-Peñarrocha, "Modal analysis and design of band-notched uwb planar monopole antennas," *IEEE Trans. Antennas Propag.*, vol. 58, no. 5, pp. 1457–1467, May 2010.

[185] E. Antonino-Daviu, M. Cabedo-Fabres, A. Valero-Nogueira, and M. Ferrando- Bataller, "A discussion on the feed configuration of planar monopole antennas to obtain ultra wide band performance," in 2004 IEEE Antennas and Propagation Society Symposium, Monterey, CA, Jul. 2004.

[186] B. J. Tomas, V. N. Alejandro, V. B. Felipe, A. D. Eva, and M. Cabedo-Fabrés, "A 60-GHz LTCC rectangular dielectric resonator antenna design with characteristic modes theory," in 2014 IEEE International Symposium on Antennas and Propagation, Memphis, TN, pp. 1928–1929, 2014.

[187] R. T. Maximidis, C. L. Zekios, T. N. Kaifas, E. E. Vafiadis, and G. A. Kyriacou, "Characteristic mode analysis of composite metal-dielectric structure based on surface integral equation/moment method," in 8th European Conference on Antennas and Propagation (EuCAP 2014), *The Hague*, the Netherlands, pp. 3418–3422, Apr. 2014.

[188] E. Safin and D. Manteuffel, "Resonance behaviour of characteristic modes due to the presence of dielectric objects," in 2013 7th European Conference on Antennas and Propagation (EUCAP), Gothenburg, Sweden, Apr. 2013.

[189] A. Krewski and W. L. Schroeder, "General application of characteristic mode equivalent circuit models in antenna design," in Proceedings of the IEEE Antennas and Propagation Society International Symposium Digest, Memphis, TN, pp. 1338–1339, Jul. 2014.

[190] K. Obeidat, B. D. Raines, and R. G. Rojas, "Discussion of series and parallel resonance phenomena in the input impedance of antennas," *Radio Sci.*, vol. 45, pp.1–9, Dec. 2010.

[191] J. J. Adams and J. T. Bernhard, "Broadband equivalent circuit models for antenna impedances and fields using characteristic modes," *IEEE Trans. Antennas Propag.*, vol. 61, no. 8, pp. 3985–3994, Aug. 2013.

[192] J. J. Adams and J. T. Bernhard, "Eigenmode-based circuit models for antennas," in Proceedings of the 2012 Antenna Applications Symposium, Monticello, IL, pp. 25–37, Sep. 2012.

[193] J. J. Adams, "Accelerated frequency interpolation of antenna impedances using characteristic modes," in Proceedings of the 2014 IEEE Antennas and Propagation International Symposium, Memphis, TN, pp. 1413–1414, Jul. 2014.

[194] Y. Chen and C.-F. Wang, "Synthesis of platform integrated antennas for reconfigurable radiation patterns using the theory of characteristic modes," in 10th International Symposium on Antennas, Propagation & EM Theory (ISAPE), Xian, China, pp. 281–285, Oct. 2012.

[195] E. Safin and D. Manteuffel, "Reconstruction of the characteristic modes on an antenna based on the radiated far field," *IEEE Trans. Antennas Propag.*, vol. 61, no. 6, pp. 2964–2971, Jun. 2013.

[196] E. Safin, R. Martens, and D. Manteuffel, "Modal source reconstruction based on radiated far-field for antenna design," in 2012 6th European Conference on Antennas and Propagation (EUCAP), Prague, Czech Republic, pp. 1645–1649, Mar. 2012.

[197] B. D. Raines and R. G. Rojas, "Wideband characteristic mode tracking," *IEEE Trans. Antennas Propag.*, vol. 60, pp. 3537–3541, 2012.

[198] B. D. Raines and R. G. Rojas, "Wideband tracking of characteristic modes," in 2011 European Conference on Antennas & Propagation (EUCAP), Rome, Italy, Apr. 11–15, 2011.

[199] D. J. Ludick, U. Jakobus, and M. Vogel, "A tracking algorithm for the eigenvectors calculated with characteristic mode analysis," in 2014 8th European Conference on Antennas and Propagation (EuCAP), Hague, the Netherlands, pp. 569–572, Apr. 2014.

[200] M. Capek, P. Hazdra, P. Hamouz, and J. Eichler, "A Method for tracking characteristic numbers and vectors," *Prog. Electromagn. Res. B*, vol. 33, pp. 115–134, 2011.

[201] O. Bucci and G. Di Massa, "Use of characteristic modes in multiple scattering problems," *J Phys. D-Appl. Phys.*, vol. 28, pp. 2235–2244, 1995.

[202] G. Amendola, G. Angiulli, and G. D. Massa, "Characteristic modes in multiple scattering by conducting cylinders of arbitrary shape," *Electromagnetics*, vol. 18, no. 6, pp. 593–612, Jun. 1998.

[203] G. Amendola, G. Angiulli, and G. D. Massa, "Scattering from arbitrarily shaped cylinders by use of characteristic modes," in Proceedings of 13th Applied Computational Electromagnetic Society Symposium, Monterey, CA, pp. 1290–1295, Mar. 1997.

[204] G. Amendola, G. Angiulli, and G. D. Massa, "Numerical and analytical characteristic modes for conducting elliptic cylinders," *Microw. Opt. Technol. Lett.*, vol. 16, no. 4, pp. 243–249, Nov. 1997.

[205] G. Amendola, G. Angiulli, and G. D. Massa, "Application of characteristic modes to the analysis of microstrip array," in Progress in Electromagnetics Research Symposium, Cambridge, MA, Jul. 2000.

[206] G. Amendola, G. Angiulli, and G. D. Massa, "Application of characteristic modes to the analysis of scattering from microstrip antennas," *J. Electromagn. Waves Appl.*, vol. 14, pp. 1063–1081, 2000.

[207] G. Angiulli and G. D. Massa, "Mutual coupling evaluation by characteristic modes in multiple scattering of electromagnetic field," in 1996 Mediterranean Electrotechnical Conference - MELECON, Bari, Italy, vol. 1, pp. 580–583, May 1996.

[208] G. Angiulli and G. D. Massa, "Radiation from arbitrary shaped microstrip antennas using the theory of characteristic modes," in Progress in Electromagnetics Research Symposium, Nantes, France, Jul. 1998.

[209] G. Angiulli and G. D. Massa, "Scattering from shaped microstrip patch antennas using the theory of characteristic modes," in IEEE International Symposium Antennas and Propagation Society, 1998, Atlanta, GA, vol. 4, pp. 1830–1833, Jun. 1998.

[210] G. Angiulli, G. Amendola and G. D. Massa, "Scattering from elliptical cylinders by use of characteristic modes," in Progress in Electromagnetics Research Symposium (PIERS 96), Innsbruck, Austria, Jul. 1996.

[211] O. M. Bucci and G. D. Massa, "Characteristic modes for mutual coupling evaluation in microstrip antennas," in 3rd International Conference on Electromagnetics in Aerospace Applications and 7th European Electromagnetic Structures Conference, Turin, Italy, Sep. 1993.

2

CHARACTERISTIC MODE THEORY FOR PEC BODIES

2.1 BACKGROUNDS

The computational electromagnetics (CEM) plays an important role in antenna engineering. It allows antenna designers to predict antenna performance at design stage. The method of moments (MoM) [1], the finite difference time domain method (FDTD) [2–4], and the finite element method (FEM) [5, 6] are the three most popular numerical techniques in the CEM. Over the years, these numerical techniques have been widely used in the analysis of many antenna and antenna arrays. With these numerical techniques, one can obtain almost all of the antenna properties such as the radiation patterns, input impedance, and radiation efficiency. They are also efficient tools for the analysis of many antenna-related electromagnetic compatibility problems. To investigate the mechanisms of the resonant behavior and the electromagnetic wave radiation from an antenna structure, antenna engineers usually have to carefully examine the current distribution on the radiating structure or look at the electric or magnetic field around the antenna structure. Because external sources and feeding structures are considered in these EM full-wave simulations, obtaining the essential physics of an antenna structure depends on whether the natural resonance has been successfully excited. In other words, an improper feeding structure design will produce misleading simulated current distributions on the antenna. These current distributions cannot truly reflect the natural electromagnetic properties of the antenna structure. Therefore, in the case that the resonant modes of an antenna are

Characteristic Modes: Theory and Applications in Antenna Engineering, First Edition.
Yikai Chen and Chao-Fu Wang.
© 2015 John Wiley & Sons, Inc. Published 2015 by John Wiley & Sons, Inc.

not properly excited, the numerical results cannot provide useful information for further optimization of the antenna and feeding structures. On the contrary, a properly designed feeding structure depends on the understanding of the natural resonant behavior of an antenna structure.

As an example, a rectangular microstrip antenna has its dominant TM_{10} mode at a specific frequency of interest. The TM_{10} mode has its own modal current distribution and modal field in the far-field zone. These modal properties are the natural behaviors of the rectangular microstrip antenna and are determined only by the geometry of the patch and the dielectric slab it resides. If we need to observe the modal properties of the TM_{10} mode through full-wave simulation, we have to properly design feeding structure to excite the TM_{10} mode. In other words, if the TM_{10} mode is not successfully excited due to an improper feeding structure, the full-wave simulation will be not able to give the current and far-field pattern for the TM_{10} mode. In most cases, however, the resonant behavior of a complicated radiation structure is usually not easy to acquire. Fortunately, the characteristic mode (CM) theory [7, 8] provides a source-free method to analyze the resonant behavior of arbitrary radiating structure. The CM analysis allows obtaining the resonant frequencies of fundamental mode as well as those higher order modes without needing any sources. It also describes the resonance behavior of each mode on the electromagnetic structure and the radiation behavior in the far-field zone. These valuable modal analysis results indicate how to excite the electromagnetic structure for achieving desired radiation performance at a specific frequency.

This chapter introduces the fundamental CM theory for perfectly electrically conducting (PEC) bodies. Traditionally, CM theory for PEC bodies is developed from the electric field integral equation (EFIE). The reason is that EFIE is applicable to both closed and open structures. Alternatively, the CM formulation for PEC bodies can be also formulated from the magnetic field integral equation (MFIE). In the following subsections, we will first formulate the EFIE and MFIE of PEC bodies from the Maxwell's equation. The MoM [1] is then applied to discretize the EFIE and MFIE into matrix equations. With the MoM impedance matrices, generalized eigenvalue equation is then developed by making use of the matrix properties, the Poynting's theorem, and the definitions for characteristic currents in the CM theory. Important CM properties, physical interpretations of many CM quantities, and their applications in antenna engineering are described in details. CM formulations based on the approximate magnetic field integral equation (AMFIE) and combined field integral equation (CFIE) are also developed to provide the flexibility of implementing the CM theory with different integral equations. All of these CM formulations provide the same CM analysis results. Following the CM theory are the algorithms for solving generalized eigenvalue equations and modal tracking across a wide frequency band. Numerical examples are finally presented to examine the numerical aspects of the CM theory and provide guidelines on how to implement the CM theory in computer code. Illustrative results are also given to show how to apply the CM theory in the analysis and design of antennas. More advanced topics on the applications of the PEC CM theory to antenna engineering will be discussed in Chapter 6.

2.2 SURFACE INTEGRAL EQUATIONS

Surface integral equation (SIE), together with the MoM, is widely used for solving time harmonic electromagnetic radiation and scattering problems from PEC objects. There are two basic SIEs for PEC structures: the electric field integral equation and the magnetic field integral equation. In the EFIE, the boundary condition is enforced on the tangential electric field, whereas in the MFIE, the boundary condition is enforced on the tangential magnetic field. We will start from Maxwell's equations and the boundary conditions to formulate the two basic SIEs.

2.2.1 Maxwell's Equations

Maxwell's equations are the fundamental equations in electromagnetics [9]. Consider an electromagnetic wave within a homogeneous medium with constituent parameters ε (permittivity) and μ (permeability), the electric and magnetic fields, electric and magnetic currents have to satisfy the following frequency-domain Maxwell's equations (a time dependence factor $e^{j\omega t}$ is suppressed):

$$\nabla \times \mathbf{E} = -\mathbf{M} - j\omega\mu\,\mathbf{H} \tag{2.1}$$

$$\nabla \times \mathbf{H} = \mathbf{J} + j\omega\varepsilon\,\mathbf{E} \tag{2.2}$$

$$\nabla \cdot \mathbf{D} = \rho_e \tag{2.3}$$

$$\nabla \cdot \mathbf{B} = \rho_m \tag{2.4}$$

where $\mathbf{E}$ is the electric field intensity, $\mathbf{H}$ is the magnetic field intensity, $\mathbf{D} = \varepsilon\mathbf{E}$ is the electric flux density, $\mathbf{B} = \mu\mathbf{H}$ is the magnetic flux density, ρ_e is the electric charge density, and ρ_m is the magnetic charge density, respectively.

In addition to the Maxwell's equations, there are another two continuity equations relating the change of the current density and the charge density:

$$\nabla \cdot \mathbf{J} = -j\omega\rho_e \tag{2.5}$$

$$\nabla \cdot \mathbf{M} = -j\omega\rho_m \tag{2.6}$$

In Equations (2.1)–(2.6), only four of the six equations are independent. We can use the two curl equations and two of the divergence equations in (2.3)–(2.6) to describe an electromagnetic problem.

2.2.2 Electromagnetic Boundary Condition

The electromagnetic fields may be discontinuous on the interface of two mediums. They are governed by the following boundary conditions:

$$\hat{n} \times (\mathbf{E}_1 - \mathbf{E}_2)\big|_s = -\mathbf{M}_s \tag{2.7}$$

$$\hat{n} \times (\mathbf{H}_1 - \mathbf{H}_2)\big|_s = \mathbf{J}_s \tag{2.8}$$

$$\hat{\mathbf{n}} \cdot (\mathbf{D}_1 - \mathbf{D}_2)\big|_s = \rho_{es} \tag{2.9}$$

$$\hat{\mathbf{n}} \cdot (\mathbf{B}_1 - \mathbf{B}_2)\big|_s = \rho_{ms} \tag{2.10}$$

where the subscripts 1 and 2 represent the medium 1 and medium 2, respectively, and $\hat{\mathbf{n}}$ is the unit normal vector on the boundary and is pointing from medium 2 to medium 1.

In particular, when the medium 2 is a perfect electric conductor, the electromagnetic fields have to satisfy the following boundary conditions:

$$\hat{\mathbf{n}} \times \mathbf{E}\big|_s = 0 \tag{2.11}$$

$$\hat{\mathbf{n}} \times \mathbf{H}\big|_s = \mathbf{J}_s \tag{2.12}$$

$$\hat{\mathbf{n}} \cdot \mathbf{D}\big|_s = \rho_{es} \tag{2.13}$$

$$\hat{\mathbf{n}} \cdot \mathbf{B}\big|_s = 0 \tag{2.14}$$

2.2.3 Magnetic Vector Potential and Electric Scalar Potential

In a source-free region, Equation (2.4) shows that $\mathbf{H}$ is always solenoidal. Therefore, it can be written as the curl of another arbitrary vector:

$$\mathbf{H} = \frac{1}{\mu} \nabla \times \mathbf{A} \tag{2.15}$$

where $\mathbf{A}$ is referred to as the magnetic vector potential. Substitution of (2.15) into (2.1) gives:

$$\nabla \times \mathbf{E} = -j\omega \nabla \times \mathbf{A} \tag{2.16}$$

Therefore, we have:

$$\nabla \times (\mathbf{E} + j\omega \mathbf{A}) = 0 \tag{2.17}$$

Applying the vector identity formulation $\nabla \times (-\nabla \Phi) = 0$, the electric field is given by:

$$\mathbf{E} = -j\omega \mathbf{A} - \nabla \Phi \tag{2.18}$$

where Φ is an arbitrary electric scalar potential. Taking the curl of both sides of Equation (2.15), and using the vector identity $\nabla \times \nabla \times \mathbf{A} = \nabla(\nabla \cdot \mathbf{A}) - \nabla^2 \mathbf{A}$, we have:

$$\mu \nabla \times \mathbf{H} = \nabla(\nabla \cdot \mathbf{A}) - \nabla^2 \mathbf{A} \tag{2.19}$$

Substituting (2.19) into (2.2), we get:

$$\nabla(\nabla \cdot \mathbf{A}) - \nabla^2 \mathbf{A} = \mu \mathbf{J} + j\omega \varepsilon \mu \mathbf{E} \tag{2.20}$$

Combining (2.18) and (2.20) leads to the following:

$$\nabla^2 \mathbf{A} + k^2 \mathbf{A} = -\mu \mathbf{J} + \nabla(\nabla \cdot \mathbf{A} + j\omega\varepsilon\,\mu\nabla\Phi) \tag{2.21}$$

where k is the wavenumber and is defined by $k = \omega\sqrt{\varepsilon\mu}$. The curl of $\mathbf{A}$ has already been defined in Equation (2.15). The Helmholtz theorem illustrates that to uniquely define the vector field $\mathbf{A}$, the divergence of $\mathbf{A}$ must be defined. Considering the right-hand side of Equation (2.21), the Lorentz gauge is adopted to simplify Equation (2.21):

$$\nabla \cdot \mathbf{A} = -j\omega\varepsilon\mu\,\Phi \tag{2.22}$$

Therefore, Equation (2.21) reduces to an inhomogeneous vector Helmholtz equation:

$$\nabla^2 \mathbf{A} + k^2 \mathbf{A} = -\mu \mathbf{J} \tag{2.23}$$

The electric field in the source-free region thus can be expressed as follows:

$$\mathbf{E} = -j\omega\mathbf{A} - \nabla\Phi = -j\omega\,\mathbf{A} - \frac{j}{\omega\varepsilon\mu}\nabla(\nabla \cdot \mathbf{A}) \tag{2.24}$$

The magnetic field can be also written in terms of the magnetic vector potential:

$$\mathbf{H} = \frac{1}{\mu}\nabla \times \mathbf{A} \tag{2.25}$$

2.2.4 Electric Field Integral Equation

Let us consider an incident plane wave $\mathbf{E}^i$ illuminating on a PEC structure. As shown in Figure 2.1, the incident wave induces surface currents $\mathbf{J}$ on the PEC body. The induced surface current $\mathbf{J}$ will then generate a scattering field $\mathbf{E}^s$. The boundary condition in (2.11) illustrates that the tangential electric field vanishes on PEC

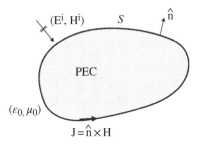

FIGURE 2.1   Surface equivalence principle of a 3D PEC structure.

surfaces S. This boundary condition gives an integral equation from which the surface current $\mathbf{J}$ can be solved:

$$(\mathbf{E}^i(\mathbf{r}) + \mathbf{E}^s(\mathbf{r}))_{\text{tan}} = 0, \quad \mathbf{r} \in S \tag{2.26}$$

where the subscript "tan" represents the tangential component of the electric field.

The scattering electric field $\mathbf{E}^s$ can be expressed in terms of the induced surface current:

$$\begin{aligned}
\mathbf{E}^s &= -j\omega\mathbf{A}(\mathbf{r}) - \nabla\Phi(\mathbf{r}) \\
&= -\frac{j\omega\mu_0}{4\pi}\int_S G(\mathbf{r},\mathbf{r}')\mathbf{J}(\mathbf{r}')dS' - \frac{j}{4\pi\varepsilon_0\omega}\nabla\int_S G(\mathbf{r},\mathbf{r}')\nabla'\cdot\mathbf{J}(\mathbf{r}')dS'
\end{aligned} \tag{2.27}$$

where ε_0 and μ_0 are the permittivity and permeability of the free space, respectively. $G(\mathbf{r},\mathbf{r}')$ is the Green's function multiplied by 4π in the free space and is given by:

$$G(\mathbf{r},\mathbf{r}') = \frac{e^{-jkR}}{R} \tag{2.28}$$

where $R = |\mathbf{r} - \mathbf{r}'|$ denotes the distance between the source point $\mathbf{r}'$ and observation point $\mathbf{r}$.

Combining the scattering field expression (2.27) with (2.26), we obtain:

$$[L(\mathbf{J})]_{\text{tan}} = \mathbf{E}^i_{\text{tan}}(\mathbf{r}), \quad \mathbf{r} \in S \tag{2.29}$$

where $L(\cdot)$ is an integro-differential operator to express the scattering field in terms of the electric current and can be written as follows:

$$L(\mathbf{J}) = -\mathbf{E}^s(\mathbf{r}) = \frac{jk_0\eta_0}{4\pi}\left(\int_S \mathbf{J}(\mathbf{r}')G(\mathbf{r},\mathbf{r}')dS' + \frac{1}{k_0^2}\nabla\int_S \nabla'\cdot\mathbf{J}(\mathbf{r}')G(\mathbf{r},\mathbf{r}')dS'\right) \tag{2.30}$$

where $k_0 = \sqrt{\varepsilon_0\mu_0}$ and $\eta_0 = \sqrt{\mu_0/\varepsilon_0}$ are the wavenumber and wave impedance in free space. Equation (2.29) is developed using the boundary condition of the electric field, and hence it is usually termed as the EFIE.

Once the induced currents $\mathbf{J}$ is solved from Equation (2.29), the scattering electric field in the far-field range can be calculated from the following:

$$\mathbf{E}^s(\mathbf{r}) = -jk_0\eta_0\frac{e^{-jkr}}{4\pi r}\int_S \mathbf{J}(\mathbf{r}')e^{jk\mathbf{r}'\cdot\hat{r}}dS' \tag{2.31}$$

where $r = |\mathbf{r}|$ and $\hat{r}$ is the unit vector in the direction of $\mathbf{r}$. The scattering magnetic field in the far-field range can be readily obtained from:

$$\mathbf{H}^s(\mathbf{r}) = \frac{1}{\eta_0}\hat{r} \times \mathbf{E}^s(\mathbf{r}) \tag{2.32}$$

2.2.5   Magnetic Field Integral Equation

Applying the boundary condition for the magnetic field as given in Equation (2.12), an MFIE for a closed and smooth PEC body can be derived:

$$\hat{\mathbf{n}} \times (\mathbf{H}^i(\mathbf{r}) + \mathbf{H}^s(\mathbf{r})) = \mathbf{J}(\mathbf{r}), \quad \mathbf{r} \in S \tag{2.33}$$

Combining the scattered magnetic field in Equation (2.25) with (2.33) and applying Cauchy principal value to the curl term, Equation (2.33) reduces to:

$$\frac{\mathbf{J}(\mathbf{r})}{2} - \hat{\mathbf{n}} \times \frac{1}{\mu} \nabla \times \tilde{\mathbf{A}}(\mathbf{r}) = \hat{\mathbf{n}} \times \mathbf{H}^i(\mathbf{r}), \quad \mathbf{r} \in S \tag{2.34}$$

where $\tilde{\mathbf{A}}(\mathbf{r})$ is the integration on the closed surface with removing the singularity at $\mathbf{r} = \mathbf{r}'$, and is given by:

$$\tilde{\mathbf{A}}(\mathbf{r}) = \frac{\mu}{4\pi} \int_{S_0} \mathbf{J}(\mathbf{r}') G(\mathbf{r}, \mathbf{r}') dS' \tag{2.35}$$

Applying the vector identity, $\nabla \times (\mathbf{J}(\mathbf{r}')G(\mathbf{r}, \mathbf{r}')) = \mathbf{J}(\mathbf{r}') \times \nabla' G(\mathbf{r}, \mathbf{r}')$, will reduce Equation (2.34) to the commonly used MFIE as follows:

$$\frac{\mathbf{J}(\mathbf{r})}{2} - \hat{\mathbf{n}} \times K(\mathbf{J}) = \hat{\mathbf{n}} \times \mathbf{H}^i(\mathbf{r}), \quad \mathbf{r} \in S \tag{2.36}$$

where the integro-differential operator $K(\mathbf{J})$ is introduced for concise representation and is given by:

$$K(\mathbf{J}) = \frac{1}{4\pi} P.V. \int_S \mathbf{J}(\mathbf{r}') \times \nabla' G(\mathbf{r}, \mathbf{r}') dS' \tag{2.37}$$

where $P.V.$ denotes the Cauchy principal value integration.

2.3   METHOD OF MOMENTS

The MoM is a way of solving an integral equation by converting it into a matrix equation. In electromagnetics, MoM is particularly well suited to open problems, such as the radiation and scattering problems [1, 10]. It obtains the matrix representation of an integral operator by expanding and approximating the unknown currents in terms of a finite set of expansion functions. The expansion function can be either entire-domain basis function or sub-domain basis function. In general, entire-domain basis functions are not available for EM problems with complex geometries. Therefore, only sub-domain basis function is adopted throughout this book.

In the MoM, there are primarily three steps in discretizing an integral equation into a matrix equation. First, the three-dimensional PEC surface will be modeled by

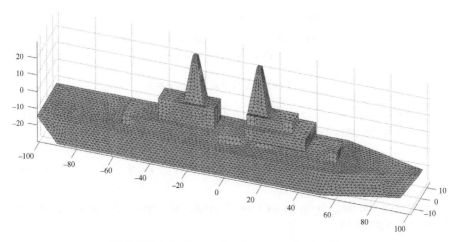

FIGURE 2.2   Triangular element mesh of a ship.

dividing the surface into many small elements. Second, appropriate basis functions should be assigned to the elements such that the unknown currents in the integral equation are expanded. Third, testing procedure is used to transfer the integral operator into a MoM impedance matrix.

In the first step, triangular elements are particularly suitable to model arbitrary shaped three-dimensional PEC surface. It is because the triangular element is a simplex in a two-dimensional surface and is easy to conformal with any curved surfaces. Many CAD softwares such as the ANSYS [11] and PATRAN [12] are able to generate high-quality triangular elements. Figure 2.2 illustrates the triangular element meshes generated by the PATRAN software for a ship model. In general, quasi-equilateral triangular elements with uniform element size are preferred to ensure the accuracy in geometry modeling and also well-conditioned MoM matrix. In general, $\lambda/10$ mesh size is usually sufficient to attain accurate results in scattering and conventional radiation problems, where λ is the wavelength in the free space. In CM analysis, however, modal currents usually change drastically in the higher order modes. In this sense, to capture the behavior of very high-order modes accurately, it is suggested to model the PEC surface with a relatively dense mesh in excess of $\lambda/15$. Numerical aspects of the influence of the mesh density on the CM analysis will be discussed in later subsections.

The next step in the MoM is to define basis functions on the triangular elements. In the method of moments, the most popular basis function for PEC surfaces is the Rao-Wilton-Glisson (RWG) basis function [13]. As shown in Figure 2.3, the RWG basis function is defined over a pair of triangular elements that share a common edge, and it is given by the following vector function:

$$\mathbf{f}(\mathbf{r}) = \mathbf{f}_n^+(\mathbf{r}) + \mathbf{f}_n^-(\mathbf{r}) \tag{2.38a}$$

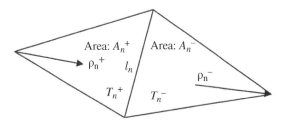

FIGURE 2.3   RWG basis function defined over a pair of adjacent triangle elements.

$$\mathbf{f}_n^{\pm}(\mathbf{r}) = \begin{cases} \dfrac{l_n}{2A_n^{\pm}} \boldsymbol{\rho}_n^{\pm}, & \mathbf{r} \in T_n^{\pm} \\ 0, & \mathbf{r} \notin T_n^{\pm} \end{cases} \tag{2.38b}$$

where T^+ and T^- are the two triangles that share a common edge which is indexed by n, l_n is the length of the edge, A_n^+ and A_n^- are the areas of the two triangles, and $\boldsymbol{\rho}_n^+$ and $\boldsymbol{\rho}_n^-$ are the position vectors defined with respect to the free vertices. In particular, $\boldsymbol{\rho}_n^+$ points away from the free vertex of T^+, and $\boldsymbol{\rho}_n^-$ points toward the free vertex of T^-. The plus and minus designations to the two triangles are based on the assumption that the positive currents are flowing from T^+ to T^-.

As observed from the RWG basis function, the normal component to the common edge l_n is a constant (can be further normalized to unity), whereas the normal component to any other edges equals to zero. This important feature ensures the current continuity across the interface of two adjacent triangles.

Moreover, taking the surface divergence of the RWG basis function in Equation (2.38b), we obtain:

$$\nabla_S \cdot \mathbf{f}_n^{\pm}(\mathbf{r}) = \begin{cases} \pm \dfrac{l_n}{A_n^{\pm}}, & \mathbf{r} \in T_n^{\pm} \\ 0, & \mathbf{r} \notin T_n^{\pm} \end{cases} \tag{2.39}$$

As have been previously described in Equation (2.5), the divergence of the current density is proportional to electric charge density. On the other hand, Equation (2.39) shows that the total charge density associated with the triangle pairs equals to zero. Therefore, there is no fictitious charge accumulated on the common edge, and the RWG basis function is known as a divergence conforming basis function.

With the RWG basis function, we can expand the electric surface currents on the PEC body as:

$$\mathbf{J}(\mathbf{r}) = \sum_{n=1}^{N} J_n \mathbf{f}_n(\mathbf{r}) \tag{2.40}$$

where J_n is the unknown weighting coefficient for the nth basis function and N denotes the number of common edges.

The last step in discretizing an integral equation into a matrix equation is to test the integral equation with expanded current representation. Substitution of Equation (2.40) into the EFIE formulation gives:

$$\frac{jk_0\eta_0}{4\pi}\left(\sum_{n=1}^{N}J_n\int_S \mathbf{f}_n(\mathbf{r}')G(\mathbf{r},\mathbf{r}')dS' + \frac{1}{k_0^2}\sum_{n=1}^{N}J_n\nabla\int_S\nabla'\cdot\mathbf{f}_n(\mathbf{r}')G(\mathbf{r},\mathbf{r}')dS'\right)_{\tan} = \mathbf{E}^i_{\tan}(\mathbf{r}) \quad (2.41)$$

Using the RWG basis function $\mathbf{f}_m(\mathbf{r})$ as the testing function and applying the vector identity $\nabla\cdot(\phi\mathbf{A}) = \mathbf{A}\cdot(\nabla\phi) + \phi(\nabla\cdot\mathbf{A})$, we obtain the following matrix equation:

$$[Z_{mn}][J_n] = [V_m] \quad (2.42)$$

where,

$$Z_{mn} = \frac{jk_0\eta_0}{4\pi}\left(A_{mn} - \frac{1}{k_0^2}B_{mn}\right) \quad (2.43)$$

$$A_{mn} = \int_S \mathbf{f}_m(\mathbf{r})\cdot\int_S \mathbf{f}_n(\mathbf{r}')G(\mathbf{r},\mathbf{r}')dS'dS \quad (2.44)$$

$$B_{mn} = \int_S \nabla_S\cdot\mathbf{f}_m(\mathbf{r})\int_S\nabla'_S\cdot\mathbf{f}_n(\mathbf{r}')G(\mathbf{r},\mathbf{r}')dS'dS \quad (2.45)$$

$$V_m = \int_S \mathbf{f}_m(\mathbf{r})\cdot\mathbf{E}^i(\mathbf{r})dS \quad (2.46)$$

In the scattering and radiation problems, we can solve the surface currents from the matrix equation in (2.42). Radar cross-section or far-field radiations can be further calculated from the surface current. For the CM analysis, we will derive a generalized eigenvalue equation from the impedance matrix $[Z_{mn}]$. The generalized eigenvalue equation is independent of any specific excitation, and thus the excitation vector $[V_m]$ is not involved in the CM analysis.

Alternatively, we can also develop the CM formulation from the MFIE given in Equation (2.36). Similar to the EFIE case, the corresponding generalized eigenvalue equation is developed from the MFIE MoM matrix. We can transform the MFIE into a matrix equation by applying the similar MoM procedure to the MFIE. The matrix equation for the MFIE is given by:

$$[Z_{mn}][J_n] = [V_m] \quad (2.47)$$

$$Z_{mn} = \frac{1}{2}\int_S \mathbf{f}_m(\mathbf{r})\cdot\mathbf{f}_n(\mathbf{r})dS - \frac{1}{4\pi}\int_S \mathbf{f}_m(\mathbf{r})\cdot\hat{\mathbf{n}}\times\int_S\mathbf{f}_n(\mathbf{r}')\times\nabla'G(\mathbf{r},\mathbf{r}')dS'dS \quad (2.48)$$

$$V_m = \int_S \mathbf{f}_m(\mathbf{r})\cdot\hat{\mathbf{n}}\times\mathbf{H}^i(\mathbf{r})dS \quad (2.49)$$

2.4 EFIE BASED CM FORMULATION

In general, CM analysis is a process of solving eigenvalue problem. In many areas, eigenvalues and eigenvectors feature certain "natural" properties of a system governed by eigenvalue equation. In mechanical engineering, for example, the eigenvalues and eigenvectors describe the vibration property of a structure. In electromagnetics, Garbacz [7] and Harrington [8] extended the eigenanalysis to radiation and scattering problems in the 1970s. They contributed a clear and concise and yet versatile theory to compute the "natural" properties of radiation and scattering problems.

This section starts with the theoretical development of the CMs, following Harrington's approach [5]. Next, another derivation begins with the Poynting's theorem which is developed to illustrate the physics described in the generalized eigenvalue equation.

2.4.1 Conventional Derivation

The EFIE presented in Equation (2.29) reveals that the $L(\cdot)$ operator builds the relationships between the electric fields and the surface currents. Because the $L(\cdot)$ operator features with the impedance property, an impedance operator $Z(\cdot)$ is defined to exhibit this impedance property:

$$Z(\mathbf{J}) = [L(\mathbf{J})]_{\text{tan}} \tag{2.50}$$

As can be seen, $Z(\cdot)$ describes tangential component of the electric field due to the induced currents $\mathbf{J}$. It is evident that the MoM matrix calculated from Equation (2.43) is a discretization form of the $Z(\cdot)$ operator. In the following developments, we use the impedance matrix $\mathbf{Z}$ to denote the EFIE MoM matrix as given in Equation (2.43).

We can write the impedance matrix $\mathbf{Z}$ in terms of its real and imaginary Hermitian parts:

$$\mathbf{Z} = \mathbf{R} + j\mathbf{X} \tag{2.51a}$$

$$\mathbf{R} = \frac{\mathbf{Z} + \mathbf{Z}^*}{2} \tag{2.51b}$$

$$\mathbf{X} = \frac{\mathbf{Z} - \mathbf{Z}^*}{2j} \tag{2.51c}$$

As observed from Equation (2.43), both $\mathbf{R}$ and $\mathbf{X}$ are real and symmetric matrices. Moreover, $\mathbf{R}$ is also a semi-definite matrix (theoretically).

Consider the following weighted eigenvalue equation:

$$\mathbf{Z}\mathbf{J}_n = \upsilon_n \mathbf{W}\mathbf{J}_n \tag{2.52}$$

where $\mathbf{J}_n$ is the eigenvector, υ_n is the eigenvalue, and $\mathbf{W}$ is a suitably chosen matrix that will diagonalize $\mathbf{Z}$. In light of the important features of the $\mathbf{R}$ and $\mathbf{X}$, if we choose

$\mathbf{W} = \mathbf{R}$, let $\upsilon_n = 1 + j\lambda_n$, and cancel the same term that appears in the left- and right-hand sides, we arrive at the following generalized eigenvalue equation:

$$\mathbf{XJ}_n = \lambda_n \mathbf{RJ}_n \tag{2.53}$$

As both $\mathbf{R}$ and $\mathbf{X}$ are symmetrical, $\mathbf{J}_n$ and λ_n are real. We will address the properties of $\mathbf{J}_n$ and λ_n in detail in the following subsections.

2.4.2 Poynting's Theorem Based Derivation

As can be observed from the conventional derivation in Section 2.4.1, the resulted generalized eigenvalue equation is dependent on the choice of the weighted matrix $\mathbf{W}$. In this sense, it is not a straightforward derivation. In this section, we will obtain the same generalized eigenvalue equation using an alternative approach. This approach starts with the well-known Poynting's theorem, and thus it is more straightforward and has clearer physical meanings.

The Poynting's theorem is considered as the law of conservation of energy in electromagnetics. In time-harmonic electromagnetics, the Poynting's theorem is given by [14]:

$$-\frac{1}{2}\iiint_V (\mathbf{H}^* \cdot \mathbf{M}_i + \mathbf{E} \cdot \mathbf{J}_i^*)\,dV = \frac{1}{2}\oiint_S (\mathbf{E} \times \mathbf{H}^*)\,dS + \frac{j\omega}{2}\iiint_V \left(\mu|\mathbf{H}|^2 - \varepsilon|\mathbf{E}|^2\right)dV$$
$$+ \frac{1}{2}\iiint_V \sigma|\mathbf{E}|^2\,dV \tag{2.54}$$

where $\mathbf{J}_i$ and $\mathbf{M}_i$ represent the electric and magnetic current sources of the field, respectively. They are commonly referred to as the impressed currents. The term in the left-hand side is the supplied power from the sources. In the right-hand side, the first term represents the power existing away from the system. In radiation problems, it is also interpreted as the radiation power. The second term is an imaginary power. It represents the stored field energy within the system. The last term is the dissipated power due to losses. In PEC problems, there is no loss material and this term equals to zero. In light of the physical meanings of each term in Equation (2.54), the Poynting's theorem states that the supplied power in an EM system must be equal to the sum of the exiting power, the stored field energy, and the dissipated power.

In the surface equivalence problem for PEC bodies, there is no magnetic current source. Together with the condition that the dissipated power equals to zero in PEC problems, Equation (2.54) reduces to:

$$-\frac{1}{2}\iiint_V \mathbf{E} \cdot \mathbf{J}_i^*\,dV = \frac{1}{2}\oiint_S (\mathbf{E} \times \mathbf{H}^*)\,dS + \frac{j\omega}{2}\iiint_V \left(\mu|\mathbf{H}|^2 - \varepsilon|\mathbf{E}|^2\right)dV \tag{2.55}$$

It is evident that, in PEC problems, the supplied power must be equal to the sum of the exiting power and stored field energy.

On the other hand, Equations (2.30) and (2.50) show that the electric field $\mathbf{E}$ generated by the electric current source $\mathbf{J}$ can be written as:

$$\mathbf{E} = -Z(\mathbf{J}) \tag{2.56}$$

Substitution of Equation (2.56) into (2.55) and represent the left-hand side term in terms of impedance matrix, we have:

$$\frac{1}{2}\left\langle \mathbf{Z}\cdot\mathbf{J},\mathbf{J}^*\right\rangle = \frac{1}{2}\left\langle \mathbf{R}\cdot\mathbf{J},\mathbf{J}^*\right\rangle + j\frac{1}{2}\left\langle \mathbf{X}\cdot\mathbf{J},\mathbf{J}^*\right\rangle$$
$$= \frac{1}{2}\oiint_S (\mathbf{E}\times\mathbf{H}^*)\,dS + \frac{j\omega}{2}\iiint_V (\mu|\mathbf{H}|^2 - \varepsilon|\mathbf{E}|^2)\,dV \qquad (2.57)$$

Because both the integrals in the right-hand side are always real, it follows that:

$$\frac{1}{2}\left\langle \mathbf{R}\cdot\mathbf{J},\mathbf{J}^*\right\rangle = \frac{1}{2}\oiint_S (\mathbf{E}\times\mathbf{H}^*)\,dS \qquad (2.58)$$

$$\frac{1}{2}\left\langle \mathbf{X}\cdot\mathbf{J},\mathbf{J}^*\right\rangle = \frac{\omega}{2}\iiint_V (\mu|\mathbf{H}|^2 - \varepsilon|\mathbf{E}|^2)\,dV \qquad (2.59)$$

Therefore, $\left\langle \mathbf{R}\cdot\mathbf{J},\mathbf{J}^*\right\rangle$ and $\left\langle \mathbf{X}\cdot\mathbf{J},\mathbf{J}^*\right\rangle$ give the radiation and stored field energy, respectively. Evidently, $\left\langle \mathbf{R}\cdot\mathbf{J},\mathbf{J}^*\right\rangle$ cannot be negative as radiated power must be positive and is always greater than zero for open structures. Therefore, the matrix $\mathbf{R}$ is a semi-definite matrix.

To achieve high-radiation efficiency in a radiation system, we expect that the radiation power can be maximized, whereas the stored energy within the radiation system can be minimized. In practical antenna designs, careful antenna structure optimization is always necessary for the purpose of enhancing the radiation capability and suppressing non-radiation modes. In mathematical representation, we can obtain high radiation efficiency by minimizing the following function:

$$f(\mathbf{J}) = \frac{P_{store}}{P_{radiation}} = \frac{\left\langle \mathbf{X}\cdot\mathbf{J},\mathbf{J}^*\right\rangle}{\left\langle \mathbf{R}\cdot\mathbf{J},\mathbf{J}^*\right\rangle} \qquad (2.60)$$

A variational expression in terms of $\mathbf{J}$ immediately results in the following generalized eigenvalue equation:

$$\mathbf{X}\mathbf{J}_n = \lambda_n \mathbf{R}\mathbf{J}_n \qquad (2.61)$$

where $\mathbf{J}_n$ and λ_n are the real eigenvectors and eigenvalues, respectively; and n is the index of the order of each mode. Equation (2.61) has the same form as the generalized eigenvalue equation in Equation (2.53).

2.4.3 Othogonality of Characteristic Modes

2.4.3.1 Othogonality of Characteristic Currents Before discussing the orthogonality of modal currents, the eigenvectors solved from Equation (2.61) are generally normalized to unit radiation power using the following criterion:

$$\left\langle \mathbf{R}\cdot\mathbf{J},\mathbf{J}^*\right\rangle = 1 \qquad (2.62)$$

With this normalization, we observe that larger electric currents in high-order modes are required to achieve the unit radiation power as in low-order modes. It this sense, the magnitude of normalized currents reveals the radiation efficiency of each mode.

The properties of the matrices $\mathbf{R}$ and $\mathbf{X}$ ensure the real eigenvectors $\mathbf{J}_n$ be orthogonal with $\mathbf{R}$ and $\mathbf{X}$. The orthogonality properties among the modal currents are given by:

$$\langle \mathbf{J}_m, \mathbf{R} \cdot \mathbf{J}_n \rangle = \langle \mathbf{J}_m^*, \mathbf{R} \cdot \mathbf{J}_n \rangle = \delta_{mn} \tag{2.63}$$

$$\langle \mathbf{J}_m, \mathbf{X} \cdot \mathbf{J}_n \rangle = \langle \mathbf{J}_m^*, \mathbf{X} \cdot \mathbf{J}_n \rangle = \lambda_n \delta_{mn} \tag{2.64}$$

$$\langle \mathbf{J}_m, \mathbf{Z} \cdot \mathbf{J}_n \rangle = \langle \mathbf{J}_m^*, \mathbf{Z} \cdot \mathbf{J}_n \rangle = (1 + j\lambda_n) \delta_{mn} \tag{2.65}$$

where,

$$\delta_{mn} = \begin{cases} 1 & m = n \\ 0 & m \neq n \end{cases} \tag{2.66}$$

Given that all eigenvectors $\mathbf{J}_n$ are real, the conjugate operation in Equations (2.63), (2.64), and (2.65) can be omitted.

In particular, the characteristic currents exhibit the orthogonality property in the following two ways:

1. The current polarizations are orthogonal with each other.
2. The current magnitudes are orthogonal with each other.

In the second case, if one mode has very strong currents at a specific point on the PEC surface, all of its orthogonal modes must have very weak modal current at the same point.

Taking advantage of the orthogonality property of modal currents, one can excite two orthogonal modes simultaneously for possible circular polarization antenna designs. It also allows the excitation of one mode and suppression of the other orthogonal modes for the enhancement of antenna's polarization purity. Moreover, the orthogonality property also provides useful information on how to properly design the feeding structures for improved isolation performance in multi-port antenna designs.

2.4.3.2 Othogonality of Characteristic Fields

The far field due to the modal currents is referred as characteristic fields. By substituting Equation (2.65) in (2.57) and omitting the common coefficients of each term, we have:

$$\langle \mathbf{J}_m^*, \mathbf{Z} \cdot \mathbf{J}_n \rangle = (1 + j\lambda_n) \delta_{mn} = \oiint_S (\mathbf{E}_m \times \mathbf{H}_n^*) dS + j\omega \iiint_V (\mu \mathbf{H}_m \cdot \mathbf{H}_n^* - \varepsilon \mathbf{E}_m \cdot \mathbf{E}_n^*) dV \tag{2.67}$$

Interchanging m and n and taking the complex conjugate of the Equation (2.67) yield:

$$(1 - j\lambda_m) \delta_{nm} = \oiint_S (\mathbf{E}_n^* \times \mathbf{H}_m) dS - j\omega \oiint_V (\mu \mathbf{H}_n^* \cdot \mathbf{H}_m - \varepsilon \mathbf{E}_n^* \cdot \mathbf{E}_m) dV \tag{2.68}$$

Because $\lambda_n \delta_{mn} = \lambda_m \delta_{nm}$, the sum of Equations (2.67) and (2.68) results in the following:

$$2\delta_{nm} = \oiint_S (\mathbf{E}_m \times \mathbf{H}_n^* + \mathbf{E}_n^* \times \mathbf{H}_m)\,dS \qquad (2.69)$$

In the far-field range, the characteristic fields are in the form of outward traveling waves:

$$\mathbf{E} = \eta \mathbf{H} \times \hat{\mathbf{k}} \qquad (2.70)$$

where $\hat{\mathbf{k}}$ is the unit vector in the wave-traveling direction. Using the relationships between $\mathbf{E}$ and $\mathbf{H}$ in Equation (2.70), Equation (2.69) can be further reduced to two equations, in terms of the electric field and magnetic field, respectively,

$$\frac{1}{\eta} \oiint_S \mathbf{E}_m \cdot \mathbf{E}_n^* \, dS = \delta_{nm} \qquad (2.71)$$

$$\eta \oiint_S \mathbf{H}_m \cdot \mathbf{H}_n^* dS = \delta_{nm} \qquad (2.72)$$

Both the equations above show that characteristic electric fields form an orthogonal set in the far field.

The characteristic fields also exhibit orthogonality property in two forms similar to that of characteristic currents:

1. The polarizations of the characteristic fields are orthogonal with each other.
2. The magnitudes of the characteristic fields are orthogonal with each other.

It is evident that features of the characteristic fields are closely related to those of the characteristic currents. A further investigation and control of the orthogonal characteristic fields are helpful in antenna designs where far-field polarization performance is of great importance. A typical application of the far-field orthogonality is in the MIMO antenna designs [15].

2.4.4 Physical Interpretation of Eigenvalues

The eigenvalue λ_n is a quantity solved immediately from the generalized eigenvalue equation. Applying Equation (2.70) in the characteristic fields, we obtain:

$$\oiint_S \mathbf{E}_m \times \mathbf{H}_n^* dS = \oiint_S \mathbf{E}_n^* \times \mathbf{H}_m dS \qquad (2.73)$$

By subtracting Equation (2.68) from (2.67) and making use of Equation (2.73), we get:

$$\omega \iiint_V (\mu \mathbf{H}_m \cdot \mathbf{H}_n^* - \varepsilon \mathbf{E}_m \cdot \mathbf{E}_n^*)\,dV = \lambda_n \delta_{mn} \qquad (2.74)$$

Choosing $m = n$ in Equation (2.74) gives us the following physical interpretations of the eigenvalues:

- The total stored field energy within a radiation or scattering problem is proportional to the magnitude of the eigenvalues.
- In the case of $\lambda_n = 0$, it illustrates that $\iiint_V \mu \mathbf{H}_n \cdot \mathbf{H}_n^* dV = \iiint_V \varepsilon \mathbf{E}_m \cdot \mathbf{E}_n^* dV$. It corresponds to the case of resonance and the associated modes are known as the resonant modes.
- In the case of $\lambda_n > 0$, it illustrates that $\iiint_V \mu \mathbf{H}_n \cdot \mathbf{H}_n^* dV > \iiint_V \varepsilon \mathbf{E}_m \cdot \mathbf{E}_n^* dV$. Because the stored magnetic field energy dominates over the stored electric field energy, the associated modes are known as the inductive modes.
- In the case of $\lambda_n < 0$, it illustrates that $\iiint_V \mu \mathbf{H}_n \cdot \mathbf{H}_n^* dV < \iiint_V \varepsilon \mathbf{E}_m \cdot \mathbf{E}_n^* dV$. Because the stored electric field energy dominates over the stored magnetic field energy, the associated modes are known as the capacitive modes.

2.4.5   Physical Interpretation of Modal Significances

CMs are a complete set of orthogonal modes for expanding any induced currents and far fields due to a specific external source. In other words, the induced currents on the PEC body can be written as a superposition of the characteristic currents:

$$\mathbf{J} = \sum_n a_n \mathbf{J}_n \tag{2.75}$$

and subsequently the far-fields produced by the induced currents can be expanded using the characteristic fields:

$$\mathbf{E} = \sum_n a_n \mathbf{E}_n \tag{2.76}$$

$$\mathbf{H} = \sum_n a_n \mathbf{E}_n \tag{2.77}$$

where a_n are the complex weighting coefficients for each mode and are to be determined. There are roughly two ways to determine the complex weighting coefficients:

1. Using the orthogonal properties of the characteristic fields.
2. Using optimization algorithms to find the optimal set of a_n such that they will give the best approximation to a designated far-field pattern.

In Chapter 6, we will discuss the determination of a_n in detail with an emphasis on the second approach.

By substituting Equation (2.75) into the EFIE in (2.29) and using the Z impedance operator denotation, we have:

$$\sum_n a_n Z(\mathbf{J}_n) = \mathbf{E}^i_{\tan}(\mathbf{r}) \tag{2.78}$$

Taking the inner product of Equation (2.78) with characteristic currents $\mathbf{J}_m$ and adopting the impedance matrix notation $\mathbf{Z}$ give:

$$\sum_n a_n \langle \mathbf{Z}\mathbf{J}_n, \mathbf{J}_m \rangle = \langle \mathbf{E}^i_{\tan}(\mathbf{r}), \mathbf{J}_m \rangle \tag{2.79}$$

Applying the orthogonality property of the characteristic currents in Equation (2.79), only the term with $m = n$ exists in the left-hand side,

$$a_n \left(1 + j\lambda_n \right) = \langle \mathbf{E}^i_{\tan}(\mathbf{r}), \mathbf{J}_n \rangle \tag{2.80}$$

which further gives

$$a_n = \frac{\langle \mathbf{E}^i_{\tan}(\mathbf{r}), \mathbf{J}_n \rangle}{1 + j\lambda_n} \tag{2.81}$$

where the inner product $\langle \mathbf{E}^i_{\tan}(\mathbf{r}), \mathbf{J}_n \rangle$ is called the modal excitation coefficient and the modal significance MS is defined as:

$$\mathrm{MS} = \left| \frac{1}{1 + j\lambda_n} \right| \tag{2.82}$$

The modal excitation coefficient states the way that the external excitation is coupled to each characteristic currents. This coupling is dependent on the position, magnitude, phase, and polarization of the external source. We can use the modal excitation coefficient as a guideline in the determination of optimal feeding locations.

The modal significance, however, is the intrinsic property of each mode and is independent of any specific external source. It states the coupling capability of each CM with external sources. Together with the modal excitation coefficient, the modal significance measures the contribution of each mode in the total electromagnetic response to a given source. As can be observed from Equation (2.82), the modal significance transforms the $[-\infty, +\infty]$ value range of eigenvalues into a much smaller range of $[0, 1]$. In many cases, it is more convenient to use the modal significance other than the eigenvalues to investigate the resonant behavior across a wide frequency band.

Furthermore, the modal significance also provides a convenient way to measure the bandwidth (BW) of each CM, particularly in the case that a specific feeding structure is not available in the initial design stage. A half-power BW can be defined according to the modal significance:

$$BW = \frac{f_H - f_L}{f_{res}} \qquad (2.83)$$

where f_{res}, f_H, and f_L are the resonant frequency, upper band frequency, and lower band frequency, respectively. They are determined from the modal significance values in a frequency band:

$$MS(f_{res}) = 1 \qquad (2.84)$$

$$MS(f_H) = MS(f_L) = \left| \frac{1}{1 + j\lambda_n} \right| = \frac{1}{\sqrt{2}} \qquad (2.85)$$

As a further extension, the modal significance is also used to identify the significant modes and non-significant modes. CMs with $MS \geq 1/\sqrt{2}$ are referred to as significant modes, whereas CMs with $MS < 1/\sqrt{2}$ are referred to as non-significant modes. In Chapter 6, we will use this criterion to determine how many significant modes should be considered in the platform antenna designs. This criterion ensures that the omitting of high-order modes will not give significant errors in the approximation of a designated radiation pattern.

2.4.6 Physical Interpretation of Characteristic Angles

The CM theory defines a set of real characteristic currents $\mathbf{J}_n$ on the surface of a PEC body. Each of the characteristic currents radiates a characteristic field $\mathbf{E}_n$ in the free-space. The tangential component of $\mathbf{E}_n$ on the PEC surface S, $\mathbf{E}_n^{tan}(S)$, is equiphase everywhere on S. Consequently, there exists a constant phase lag between the real current $\mathbf{J}_n$ and the equiphase $\mathbf{E}_n^{tan}(S)$. Because $\mathbf{E}_n^{tan}(S)$ can be written in terms of the impedance operator, the constant phase lag between $\mathbf{E}_n^{tan}(S)$ and the real current $\mathbf{J}_n$ can be calculated from [16]:

$$\alpha_n = 180° - \tan^{-1}\lambda_n \qquad (2.86)$$

As can be seen, this phase lag can be computed from the eigenvalue immediately. In CM theory, it is usually called characteristic angle. It provides a way to better show the mode behavior near resonance. If the modal current and tangential electric field on S are 180° out of phase ($\alpha_n = 180°$), the mode is said to be in external (scattering or radiation) resonance and the PEC body is a most effective radiator. On the contrary, if the modal current and the tangential electric field on S are 90° or 270° out of phase ($\alpha_n = 90°$ or 270°), the mode is said to be in internal (cavity) resonance. In this case, the modal current yields a null field in the exterior region. Therefore, the variation of the characteristic angle with frequency provides qualitative information to describe the radiation or scattering capability of each mode. In particular, the phase information in the characteristic angle allows us to determine whether two orthogonal modes can be excited with one feeding for circular polarization antenna designs. In this case, two orthogonal modes with comparable modal significance and 90° characteristic angle difference are required for satisfactory axial ratio performance. We will further address this topic in Chapter 3.

More often than not, the characteristic angle varies in the range [90°, 270°]. Following the physical interpretations of eigenvalues, the characteristic modes can be also categorized using the characteristic angles:

- In the case of $\alpha_n = 180°$, the associated modes are resonant modes.
- In the case of $90° < \alpha_n < 180°$, the associated modes are inductive modes.
- In the case of $180° < \alpha_n < 270°$, the associated modes are capacitive modes.

2.5   MFIE BASED CM FORMULATION

This section presents two different derivations of the CM formulations based on MFIE and AMFIE. As expected, these two methods lead to the same generalized eigenvalue equation. The first approach derives the CM formulation by following the CM definition that the characteristic current on PEC body has to be real. In the second approach, the CM formulation is derived from the AMFIE equation. The adjoint of the MFIE operator and the resultant symmetric MoM matrix for the AMFIE is applied in a derivation procedure similar to the EFIE case.

2.5.1   MFIE Based CM Formulation

Nalbantoglu proposed an alternative approach to computing the CMs for PEC bodies [17]. The CM formulation was developed from MFIE. The motivation for this new CM formulation development is to mitigate the possible ill-conditional problem in the EFIE-based MoM matrix, although this is usually not a severe problem in most CM analyses. Numerous numerical implementations have demonstrated that the EFIE-based CM formulation is free of this issue in a variety of antenna design problems. This may be due to the fact that typical radiation problems, except for those antenna arrays or reflector antennas, are in electrical small or medium sizes. Fortunately, CM theory is intended for the analysis of electromagnetic problems with electrically small and medium sizes. For electrically large problems, there are too many closely spaced resonances to identify using the CM theory. Therefore, the CM theory is not so popular in modal analysis of electrically large problems.

This section describes the development of CM formulation from the MFIE for PEC bodies. For convenience purpose, we rewrite the magnetic field equation in the following:

$$\frac{\mathbf{J}(\mathbf{r})}{2} = \hat{\mathbf{n}} \times \mathbf{H}^i(\mathbf{r}) + \hat{\mathbf{n}} \times K(\mathbf{J}), \quad \mathbf{r} \in S \tag{2.87}$$

It shows that $\hat{\mathbf{n}} \times K(\mathbf{J})$ is proportional to the tangential component of the scattered magnetic field on S. In the characteristic mode theory, it assumes that the characteristic currents $\mathbf{J}_n$ have to be real on S. Letting $\mathbf{H}^s(\mathbf{r})$ be the scattered magnetic field on the PEC surface due to $\mathbf{J}_n$, we can find that $\mathbf{H}^s(\mathbf{r}) - \text{conj}\left(\mathbf{H}^s(\mathbf{r})\right)$ is a standing wave satisfying the source-free wave equation. Because the characteristic currents are real, subtracting the conjugate of Equation (2.87) from itself yields the following:

$$0 = 2\hat{\mathbf{n}} \times \mathbf{H}^i(\mathbf{r}) + \hat{\mathbf{n}} \times K(\mathbf{J}) - \text{conj}\left(\hat{\mathbf{n}} \times K(\mathbf{J})\right) = 2\hat{\mathbf{n}} \times \mathbf{H}^i(\mathbf{r}) + 2K_I(\mathbf{J}) \qquad (2.88)$$

where $K_I(\mathbf{J})$ is the imaginary Hermitian part of the $\hat{\mathbf{n}} \times K(\mathbf{J})$. To seek a complete set of orthogonal currents that are satisfying Equation (2.88), an eigenvalue equation in terms of $K_I(\mathbf{J})$ and $\mathbf{J}$ can be constructed:

$$\hat{\mathbf{n}} \times \mathbf{H}^i(\mathbf{r}) = \gamma K_I(\mathbf{J}) \qquad (2.89)$$

where $\gamma = \rho + j\beta$ is a complex constant. Substitution of Equation (2.89) into (2.87) results in:

$$\frac{\mathbf{J}(\mathbf{r})}{2} = (\rho + j\beta)K_I(\mathbf{J}) + \hat{\mathbf{n}} \times K(\mathbf{J}) \qquad (2.90)$$

To ensure that the characteristic currents in (2.90) is real, we must have $\beta = -1$. By applying $\beta = -1$ in (2.90), we obtain the following generalized eigenvalue equation:

$$\frac{\mathbf{J}(\mathbf{r})}{2} - K_R(\mathbf{J}) = \rho K_I(\mathbf{J}) \qquad (2.91)$$

where $K_R(\mathbf{J})$ is the real Hermitian part of the $\hat{\mathbf{n}} \times K(\mathbf{J})$.

The MoM matrix element of the MFIE is given in Equation (2.48). For the sake of convenience, we also use $\mathbf{Z} = \mathbf{R} + j\mathbf{X}$ to denote it, where $\mathbf{R}$ and $\mathbf{X}$ are the real and imaginary Hermitian parts of $\mathbf{Z}$, respectively. Compare Equation (2.91) with Equation (2.48), we obtain the discretization form of the above generalized eigenvalue equation:

$$\mathbf{R}\mathbf{J}_n = -\rho_n \mathbf{X}\mathbf{J}_n \qquad (2.92a)$$

$$\mathbf{R} = \frac{\mathbf{Z} + \mathbf{Z}^*}{2} \qquad (2.92b)$$

$$\mathbf{X} = \frac{\mathbf{Z} - \mathbf{Z}^*}{2j} \qquad (2.92c)$$

To be consistent with the generalized eigenvalue equation based on the EFIE, letting $\lambda_n = -\rho_n$ can finally arrive at the generalized eigenvalue equation based on the MFIE:

$$\mathbf{R}\mathbf{J}_n = \lambda_n \mathbf{X}\mathbf{J}_n \qquad (2.93)$$

The generalized eigenvalue equations based on the EFIE and MFIE result in the same set of CMs. Numerical results will be presented in the following subsections.

2.5.2 Approximate MFIE-Based CM Formulation

2.5.2.1 Approximate Magnetic Field Integral Equation
As can be observed from the MoM matrix shown in Equation (2.48), the first term is evidently symmetric, and the second term is non-symmetric, which is caused by the nonself adjoint operator in the MFIE. In addition, the MFIE solution in the exterior region is non-unique at the interior or cavity resonant frequencies of a PEC body. This problem is often referred to as the interior resonance problem. In Refs. [18] and [19], Correia and Barbosa proposed an approximate magnetic field integral equation to remove the interior resonance that occurred at the resonance frequencies. The interior resonance problem also exists in the EFIE, and correspondingly the approximate electric field integral equation (AEFIE) is developed to deal with the non-unique solution at resonant frequencies [20]. Because the resultant MoM matrix of the AMFIE is a symmetric and Hermitian matrix, the CPU time and memory are reduced to about half of those required in MFIE.

Let us introduce a new operator $L_H[J(r)] = -\hat{n} \times K[J(r)]$ to denote the non-self-adjoint operator in the MFIE. With this operator, the MFIE in Equation (2.36) can be rewritten as follows:

$$\frac{1}{2}J(r) + L_H[J(r)] = \hat{n} \times H^i(r), \quad r \in S \tag{2.94}$$

Numerical examination for the MFIE matrix confirms that the diagonal elements are at least one order of magnitude larger than the non-diagonal elements. It is thus reasonable to assume that the matrix should not be greatly changed by averaging the non-diagonal elements with their adjoint counterparts [18]. The adjoint of the term that contributes to the non-diagonal elements is given by Ref. [18]:

$$L_H^A[J(r)] = \frac{1}{4\pi} P.V. \int_S [\hat{n} \times J(r')] \times \nabla' G(r,r') dS' \tag{2.95}$$

The AMFIE is thus obtained by averaging the $L_H[J(r)]$ operator (contributes to non-diagonal elements in MoM matrix) in Equation (2.94) with its adjoint operator $L_H^A[J(r)]$:

$$\frac{1}{2}J(r) - \hat{n} \times \frac{1}{8\pi} P.V. \int_S J(r') \times \nabla' G(r,r') dS'$$
$$+ \frac{1}{8\pi} P.V. \int_S [\hat{n} \times J(r')] \times \nabla' G(r,r') dS' = \hat{n} \times H^i(r), \quad r \in S \tag{2.96}$$

Applying Galerkin's method to Equation (2.96), the MoM matrix elements for AMFIE are obtained:

$$Z_{mn}^{\text{AMFIE}} = \frac{1}{2}\langle \mathbf{f}_m(\mathbf{r}), \mathbf{f}_n(\mathbf{r})\rangle + \frac{1}{2}\left(\langle \mathbf{f}_m(\mathbf{r}), L_{\text{H}}\left[\mathbf{f}_n(\mathbf{r}')\right]\rangle + \langle \mathbf{f}_m(\mathbf{r}), L_{\text{H}}^{\text{A}}\left[\mathbf{f}_n(\mathbf{r}')\right]\rangle\right) \quad (2.97)$$

where $\mathbf{f}_n(\mathbf{r}')$ and $\mathbf{f}_m(\mathbf{r})$ are the RWG basis function and testing function in the Galerkin's procedure, respectively. Using the properties of the adjoint operator and the symmetry of the inner product, the MoM matrix elements can be expressed as:

$$Z_{mn}^{\text{AMFIE}} = \frac{1}{2}\langle \mathbf{f}_m(\mathbf{r}), \mathbf{f}_n(\mathbf{r})\rangle + \frac{1}{2}\left[\langle \mathbf{f}_m(\mathbf{r}), L_{\text{H}}\left[\mathbf{f}_n(\mathbf{r}')\right]\rangle + \langle \mathbf{f}_n(\mathbf{r}), L_{\text{H}}\left[\mathbf{f}_m(\mathbf{r}')\right]\rangle\right] \quad (2.98)$$

As can be observed from Equation (2.98), it is not necessary to evaluate the inner product explicitly for the adjoint term in the MoM matrix evaluation. Only the inner product for the $L_{\text{H}}[\mathbf{f}(\mathbf{r})]$ operator needs to be evaluated. Therefore, there is no increase in CPU time for the MoM matrix evaluation for AMFIE.

2.5.2.2 AMFIE Based Generalized Eigenvalue Equation The resultant MoM matrix given by Equation (2.98) is a symmetric and Hermitian matrix. It can be expressed in terms of its real and imaginary parts:

$$\mathbf{Z}^{\text{AMFIE}} = \mathbf{R} + j\mathbf{X} \quad (2.99)$$

where $\mathbf{R}$ and $\mathbf{X}$ are the real and imaginary parts of $\mathbf{Z}^{\text{AMFIE}}$. Apparently, both $\mathbf{R}$ and $\mathbf{X}$ are real and symmetric matrices. Consider the following weighted eigenvalue equation for $\mathbf{Z}^{\text{AMFIE}}$:

$$\mathbf{Z}^{\text{AMFIE}}\mathbf{J}_n = \upsilon_n \mathbf{W}\mathbf{J}_n \quad (2.100)$$

where $\mathbf{J}_n$ is the eigenvector, υ_n is the eigenvalue, and $\mathbf{W}$ is a suitably chosen matrix that will diagonalize $\mathbf{Z}$. Choosing $\mathbf{W} = \mathbf{X}$, letting $\upsilon_n = \lambda_n + j$, and cancelling the same term appears in the left- and right-hand sides, we can get the following generalized eigenvalue equation:

$$\mathbf{R}\mathbf{J}_n = \lambda_n \mathbf{X}\mathbf{J}_n \quad (2.101)$$

As both $\mathbf{R}$ and $\mathbf{X}$ are symmetric, $\mathbf{J}_n$ and λ_n are real. This generalized eigenvalue equation is the same as that obtained from the MFIE. The difference between the two derivations is that in the AMFIE the averaging operation is performed to the non-self-adjoint operator, whereas in the MFIE the averaging operation is directly applied to the MoM matrix. However, the two derivations are equivalent to each other, and the resulting characteristic mode solutions from Equations (2.93) and (2.101) are the same.

2.6 CFIE-BASED CM FORMULATION

The EFIE and MFIE are developed by only enforcing the boundary conditions for the tangential electric and magnetic fields, respectively. Both the EFIE and MFIE suffer from interior resonance problem at resonant frequency as discussed in [20]. As a result, interior resonance takes place in the EFIE and MFIE at the resonant frequencies of the cavity formed by the boundary of a closed PEC body. In addition to the AMFIE and AEFIE [20], another possible way to obtain unique solution at internal resonant frequency is to linearly combine the EFIE and MFIE. The resultant integral equation is referred to as the combined field integral equation, and is commonly expressed as:

$$\alpha\left(-\mathbf{E}^{s}(\mathbf{r})\right)_{\tan} + (1-\alpha)\eta_0\left(\frac{\mathbf{J}(\mathbf{r})}{2} - \hat{\mathbf{n}} \times K(\mathbf{J})\right)$$
$$= \alpha\left(\mathbf{E}^{i}(\mathbf{r})\right)_{\tan} + (1-\alpha)\eta_0 \cdot \hat{\mathbf{n}} \times \mathbf{H}^{i}(\mathbf{r}), \quad \mathbf{r} \in S \tag{2.102}$$

where $\mathbf{E}^{s}(\mathbf{r})$ and $K(\mathbf{J})$ are defined in Equations (2.27) and (2.37), respectively. α is the combination coefficient, and its value varied between 0 (MFIE) and 1 (EFIE). η_0 is the wave impedance in free space. In the CM formulation development, however, the CFIE has to be constructed in a different form to allow the diagonalization of the EFIE and MFIE contributions with their real and imaginary Hermitian parts, respectively. The CFIE for CM formulation is given by:

$$\alpha\left(-\mathbf{E}^{s}(\mathbf{r})\right)_{\tan} + (1-\alpha)j\eta_0\left(\frac{\mathbf{J}(\mathbf{r})}{2} - \hat{\mathbf{n}} \times K(\mathbf{J})\right)$$
$$= \alpha\left(\mathbf{E}^{i}(\mathbf{r})\right)_{\tan} + (1-\alpha)j\eta_0 \cdot \hat{\mathbf{n}} \times \mathbf{H}^{i}(\mathbf{r}), \quad \mathbf{r} \in S \tag{2.103}$$

The CFIE-based generalized eigenvalue equation for CM can be developed from the MoM matrix $\mathbf{Z}^{\text{CFIE}}$ for the CFIE in Equation (2.103) in a similar way as in the EFIE and MFIE cases:

$$\mathbf{X}\mathbf{J}_n = \lambda_n \mathbf{R}\mathbf{J}_n \tag{2.104}$$

where $\mathbf{Z}^{\text{CFIE}} = \mathbf{R} + j\mathbf{X}$, $\mathbf{R}$ and $\mathbf{X}$ are the real and imaginary Hermitian parts of $\mathbf{Z}^{\text{CFIE}}$, respectively; and $\mathbf{J}_n$ and λ_n are the real characteristic currents and the associated eigenvalues, respectively. The CMs solved from Equation (2.104) are the same as those solved from its EFIE and MFIE counterparts. Numerical results from the CFIE-based CM formulation will be presented in the following subsections. The effect of the combination coefficient α on the solved characteristic modes will also be investigated.

2.7 APPLICABILITY OF THE CM FORMULATIONS

2.7.1 Closed and Open Objects

In general, PEC objects can be categorized into two types: closed objects and open objects. The surface of a closed object encloses a nonzero volume, whereas the surface of an open PEC body does not enclose any volume. For example, a PEC sphere is a closed object because it has non-zero volume, and a parabolic reflector made of infinite thin PEC sheet can be taken as an open object. Figure 2.4 shows the geometries of the two examples.

The EFIE given in Equation (2.29) is suitable for both open and closed objects. However, the MFIE in Equation (2.36) is obtained using the extinction theorem of closed surfaces. Therefore, the MFIE is only suitable for closed objects. In the same sense, the CM formulation derived from the EFIE can be applied to both closed and open objects, whereas the CM formulation derived from the MFIE is only suitable to closed objects. A literature survey shows that the CM formulation based on the EFIE is more popular than that derived from the MFIE. Numerical aspects of the two CM formulations will also be investigated to show the EFIE CM formulation outperforms its MFIE counterpart in terms of numerical accuracy and stability. Moreover, most of antenna structures are open objects. This is also an important factor that leads to the popularity of the EFIE-based CM formulation.

2.7.2 Electrically Small and Large Problems

One of the most attractive features of the CM theory is that it can find the resonant frequencies of fundamental modes as well as those high-order modes. On the other hand, the electrical size of an electromagnetic object directly determines the resonant frequencies. In general, the CM theory is only applied in electromagnetic problems with electrically small and medium sizes. In these problems, only a few modes will be the dominant modes at a specific frequency. The CM theory is able to identify these different resonant modes and their resonant frequencies. However, in electrically large problems, there are many closely spaced resonant modes, even in a narrow frequency band. In this case, it is usually impossible to excite a single mode.

(a)   (b)

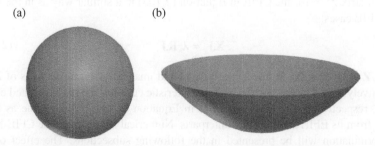

FIGURE 2.4 Example of closed and open objects. (a) Closed sphere and (b) open surface of a parabolic reflector.

Any external source will excite more than one mode. Therefore, the CM analysis will not be so helpful in the analysis of antenna structures of large electrical size. This is also the reason that the CM theory is not so popular in the electrical large problems. Therefore, most of the CM-related antenna designs focus on those with electrically small and medium sizes.

2.8 COMPUTATION OF CHARACTERISTIC MODES

So far, we have described the complete CM theory for PEC objects. In this section, we will look at the numerical implementation of the CM theory in computer code. In general, we are interested in the following three groups of quantities in the CM analysis:

1. CM parameters: eigenvalue, modal significance, and characteristic angle
2. Characteristic currents
3. Characteristic fields, including near fields and far fields

To better understand the key modules in a typical CM analysis code, Figure 2.5 shows the flowchart of a typical CM code. It can be roughly divided into five modules as shown in the first level of the flowchart,

1. Read geometry information and control parameters such as frequencies and number of modes
2. Calculate the MoM matrix from the integral equation
3. Build the generalized eigenvalue equation from the MoM matrix
4. Solve the generalized eigenvalue equation
5. Calculate the characteristic currents and characteristic fields

The second and third levels detail the computations in the above modules. In the following subsections, we will investigate two of the most important computational modules, that is, solving of the generalized eigenvalue equation and CM tracking. Other computational modules can be implemented immediately from the aforementioned formulations. Without loss of generality, the discussions in the following subsections are based on the generalized eigenvalue equation for the EFIE. In the numerical implementation of the CM formulation based on MFIE, only a slight modification is required to the one for the EFIE.

2.8.1 Solution of Generalized Eigenvalue Equation

This section considers the numerical solution of the eigenvalues λ_n and its associated eigenvectors $\mathbf{J}_n$ from the generalized eigenvalue problem (GEP) in (2.53). Harrington and Mautz [21] proposed a method using the singular value decomposition (SVD) of the matrix $\mathbf{R}$. A standard eigenvalue equation is then derived by using the

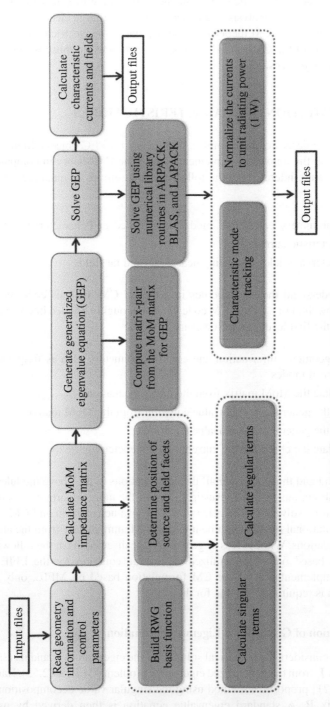

FIGURE 2.5 Flowchart of the CM computer code.

properties of the orthogonal matrices from the SVD. This approach is more suitable for problems with very small number of unknowns. On the other hand, it is known that the number of eigenvalues and eigenvectors is equal to the number of MoM unknowns. Therefore, there are a large number of eigenvalues for problems having large number of MoM unknowns. However, we are only interested in a few number of low-order modes in radiation and scattering problems. These low-order modes are sufficient to describe the resonant behaviors. Therefore, it is more preferable to compute the first several eigenvalues by using iterative methods such as the Arnoldi and Lanczos algorithm. The iterative methods will converge to a subset of the eigenpairs by repeating the matrix-vector multiplication. We are able to implement the implicitly restarted Arnoldi method using the ARPACK library [22] in the CM code. In the following subsections, we will give the details of the implementations of Harrington and Mautz's method and the implicitly restarted Arnoldi method in the CM code.

2.8.1.1 Harrington and Mautz's Method Under the assumption that there are N MoM unknowns defined for a PEC object, both matrices $\mathbf{R}$ and $\mathbf{X}$ are $N \times N$ matrices. Following Harrington and Mautz's method in Ref. [21], the computation starts with the SVD of $\mathbf{R}$:

$$\mathbf{R} = \mathbf{U}\mathbf{W}\mathbf{V}^{\mathrm{T}} \tag{2.105}$$

where $\mathbf{U}$ is an $N \times N$ unitary matrix ($\mathbf{U}\mathbf{U}^{\mathrm{T}} = \mathbf{I}$), $\mathbf{W}$ is an $N \times N$ diagonal matrix with non-negative real numbers on the diagonal, and $\mathbf{V}^{\mathrm{T}}$ denotes the transpose of an $N \times N$ unitary matrix $\mathbf{V}$ ($\mathbf{V}\mathbf{V}^{\mathrm{T}} = \mathbf{I}$). The diagonal entries σ_i of $\mathbf{W}$ are known as the singular values of $\mathbf{R}$. They are uniquely determined by $\mathbf{R}$. Conventionally, the matrix $\mathbf{W}$ is ordered according to the decreasing of singular values. Since both $\mathbf{U}$ and $\mathbf{V}^{\mathrm{T}}$ are unitary, their columns form a set of orthonormal vectors. Therefore, $\mathbf{U}$ and $\mathbf{V}^{\mathrm{T}}$ are also known as orthogonal matrices.

Using the properties of the orthogonal matrices, the GEP in Equation (2.53) can be rewritten as:

$$\mathbf{U}^{\mathrm{T}}\mathbf{X}\mathbf{V}\left(\mathbf{V}^{\mathrm{T}}\mathbf{J}_n\right) = \lambda_n \mathbf{W}\left(\mathbf{V}^{\mathrm{T}}\mathbf{J}_n\right) \tag{2.106}$$

The singular values in $\mathbf{W}$ are set to zero if they are less than a given threshold. As suggested in Ref. [21], if a singular value is smaller than one thousandth of the largest singular value, this singular value can be set to zero without introducing significant errors. Following this guideline, we assume there are N_1 non-zero singular values left in the diagonal matrix $\mathbf{W}$ ($N_1 < N$). Then, Equation (2.106) can be partitioned into the following form according to N_1:

$$\mathbf{W} = \begin{bmatrix} \mathbf{W}_{11} & 0 \\ 0 & 0 \end{bmatrix} \tag{2.107}$$

$$\mathbf{B} = \mathbf{V}^{\mathrm{T}} \mathbf{J}_n = \begin{bmatrix} \mathbf{B}_1 \\ \mathbf{B}_2 \end{bmatrix} \qquad (2.108)$$

$$\mathbf{A} = \mathbf{U}^{\mathrm{T}} \mathbf{X} \mathbf{V} = \begin{bmatrix} \mathbf{A}_{11} & \mathbf{A}_{12} \\ \mathbf{A}_{12}^{\mathrm{T}} & \mathbf{A}_{22} \end{bmatrix} \qquad (2.109)$$

Substituting Equations (2.107)–(2.109) into (2.106), we obtain the following two matrix equations:

$$\mathbf{A}_{11} \mathbf{B}_1 + \mathbf{A}_{12} \mathbf{B}_2 = \lambda_n \mathbf{W}_{11} \mathbf{B}_1 \qquad (2.110)$$

$$\mathbf{A}_{12}^{\mathrm{T}} \mathbf{B}_1 + \mathbf{A}_{22} \mathbf{B}_2 = 0 \qquad (2.111)$$

Solving $\mathbf{B}_2$ from (2.111) and substituting into (2.110) yields:

$$\left[\mathbf{A}_{11} - \mathbf{A}_{12} \mathbf{A}_{22}^{-1} \mathbf{A}_{12}^{\mathrm{T}} \right] \mathbf{B}_1 = \lambda_n \mathbf{W}_{11} \mathbf{B}_1 \qquad (2.112)$$

Because $\mathbf{W}_{11}$ is a diagonal matrix, it can be written as $\mathbf{W}_{11} = \mathbf{W}_{11}^{1/2} \cdot \mathbf{W}_{11}^{1/2}$. Substituting it into Equation (2.112) and multiplying by $\mathbf{W}_{11}^{-1/2}$, Equation (2.112) can be further reduced to a real symmetric unweighted eigenvalue equation:

$$\mathbf{C} \mathbf{Y} = \lambda_n \mathbf{Y} \qquad (2.113)$$

where,

$$\mathbf{C} = \mathbf{W}_{11}^{-1/2} \left[\mathbf{A}_{11} - \mathbf{A}_{12} \mathbf{A}_{22}^{-1} \mathbf{A}_{12}^{\mathrm{T}} \right] \mathbf{W}_{11}^{-1/2} \qquad (2.114)$$

$$\mathbf{Y} = \mathbf{W}_{11}^{1/2} \mathbf{B}_1 \qquad (2.115)$$

Eigenvalues of Equation (2.112) provide the first N_1 smallest eigenvalues of the GEP in Equation (2.53). Once the eigenvalue equation in (2.113) is solved, the eigenvectors can be readily calculated using Equations (2.115) and (2.108).

2.8.1.2 *Implicitly Restarted Arnoldi Method*

The implicitly restarted Arnoldi method (IRAM) is a robust and efficient iterative method for the numerical solution of a generalized eigenvalue equation. Fortran subroutine library that implements many IRAM variants is available in a free numerical library called ARPACK [22]. The ARPACK is able to compute a small set of eigenvalues with user specified features including smallest eigenvalues and largest eigenvalues. It is capable of solving a great variety of large-scale generalized eigenproblems from single precision positive definite symmetric problems to double precision complex non-Hermitian generalized problems. As the ARPACK library has been integrated into many software packages [22], the accuracy and stability is tested in many projects over the years. Therefore, we will discuss the algorithms of IRAM deeply and also how to implement the ARPACK in CM analysis. Readers can refer Ref. [23] for more mathematical details of the IRAM.

The first step in the implementation of IRAM is to reduce the GEP into a standard eigenvalue equation. For convenience, we rewrite the GEP of the EFIE in the following:

$$\mathbf{X}\mathbf{J}_n = \lambda_n \mathbf{R}\mathbf{J}_n \qquad (2.116)$$

Investigations to the integral operator of EFIE find that $\mathbf{X}$ is usually not ill-conditioned, except at those resonant complex frequencies. However, the CM analysis is always conducted along the real frequency axis. Another important consideration is that CM analysis is usually performed for electrical small and medium problems. Therefore, we can find the inverse of $\mathbf{X}$ without any difficulty by using typical numerical techniques such as the lower-upper (LU) decomposition. Eventually, Equation (2.116) is transformed into a standard eigenvalue equation:

$$\mathbf{X}^{-1}\mathbf{R}\mathbf{J}_n = \frac{1}{\lambda_n}\mathbf{J}_n \qquad (2.117)$$

The algorithm for this transformation is implemented by using the LAPACK routine SGETRF [24]. We can then implement the IRAM algorithm to $\mathbf{X}^{-1}\mathbf{R}$ using the ARPACK routine SNAUPD. Because we are only interested in CMs with small eigenvalue magnitude, the eigenvalue preference we choose for solving Equation (2.117) should be those with largest magnitude. Therefore, the input parameter "which" for SNAUPD should be set to "LM". A template showing how to call the SNAUPD repeatedly to solve the standard eigenvalue problem is given in the ARPACK routine SNSIMP. Here, the first letter "S" in the routine names indicates that single precision is used in the computation. One can also implement the IRAM algorithm for double precision, single precision complex, and double precision complex eigenvalue problems using ARPACK routines DNAUPD, CNAUPD, and ZNAUPD, respectively.

The most time consuming step is the LU decomposition for matrix $\mathbf{X}$. The computational time is on the order of N^3 and N is the number of MoM unknowns. In the extreme case that when electromagnetic problems involving large number of MoM unknowns need CM analysis, the LU decomposition is quite time consuming. However, we may not encounter such large problems in practice because the CM theory is more suitable to electrically small and medium problems. Moreover, even if we really need to do CM analysis for large problems, we can make use of parallel and highly optimized LU implementations. These techniques have been developed and widely used in the field of CEMs.

2.8.2   Characteristic Mode Tracking

The CM theory is developed from integral equations in frequency domain. If we are interested in the wideband response of the CMs, we have to perform the CM analysis repeatedly at many sampling frequencies. More often than not, we will find that the obtained CMs with the same mode index at different frequencies may not be the same physical mode. Therefore, CM tracking algorithms, which will track many modes across a wide frequency range, is proposed. The benefit of the mode tracking is that it makes the modes with the same index number at different frequencies be the same mode. Here, the same mode refers to the mode that has similar current distribution and modal field patterns. CM tracking gives a way to investigate how the mode evolves from non-resonant frequencies to resonant frequencies.

Two CM tracking algorithms have been reported in Refs. [25], [26]. This section presents a simple and computational efficient CM tracking algorithm. A new correlation coefficient matrix is proposed to identify the eigenvector correlations at different frequencies. It does not require the matrix inversion in the computation of correlation coefficient matrix as in Ref. [26]. This new CM tracking algorithm has been implemented to a multitude of CM analyses and has shown that it can track a large number of modes over a wide frequency band.

2.8.2.1 Key Steps There are three key steps in the tracking of a set of modes to those at another frequency. Assume that $[\mathbf{J}_1(\omega_1), \mathbf{J}_2(\omega_1), ..., \mathbf{J}_N(\omega_1)]$ and $[\mathbf{J}_1(\omega_2), \mathbf{J}_2(\omega_2), ..., \mathbf{J}_N(\omega_2)]$ are N eigenvectors at the angular frequencies ω_1 and ω_2, respectively.

In the first step, all the eigenvectors should be normalized to ensure they satisfy the following formulations:

$$\left\langle \mathbf{J}_i^{\mathrm{T}}(\omega_1), \mathbf{R}(\omega_1) \cdot \mathbf{J}_i(\omega_1) \right\rangle = 1 \tag{2.118}$$

$$\left\langle \mathbf{J}_i^{\mathrm{T}}(\omega_2), \mathbf{R}(\omega_2) \cdot \mathbf{J}_i(\omega_2) \right\rangle = 1 \tag{2.119}$$

The eigenvector normalization is mandatory in the CM tracking. It ensures all the eigenvectors are "**R**-orthogonal" and give unit radiation power.

The second step is to compute the correlation coefficient matrix $\mathbf{C} = [c_{ij}]_{N \times N}$ for the modes at two different frequencies. The correlation coefficient matrix is defined to relate the eigenvector at ω_1 to that at ω_2. The element in the correlation coefficient matrix is defined as:

$$c_{ij} = abs\left(\rho\left(\mathbf{J}_i(\omega_1), \mathbf{J}_j(\omega_2) \right) \right) \tag{2.120}$$

where $\rho(\mathbf{J}_1, \mathbf{J}_2)$ is the correlation coefficient between two eigenvectors. The correlation coefficient measures the strength and direction of a linear relationship between two vectors, and is defined by,

$$\rho(\mathbf{x}, \mathbf{y}) = \frac{N \sum_{i=1}^{N} x_i y_i - \sum_{i=1}^{N} x_i \cdot \sum_{i=1}^{N} y_i}{\sqrt{N \sum_{i=1}^{N} x_i^2 - \left(\sum_{i=1}^{N} x_i \right)^2} \cdot \sqrt{N \sum_{i=1}^{N} y_i^2 - \left(\sum_{i=1}^{N} y_i \right)^2}} \tag{2.121}$$

where x_i and y_i are the elements in vectors $\mathbf{x}$ and $\mathbf{y}$, respectively. N is the length of the vectors.

The value of ρ is in the range $-1 \le \rho \le 1$. The signs "+" and "–" denote the positive linear correlations and negative linear correlations, respectively. If $\mathbf{x}$ and $\mathbf{y}$ have a strong positive linear correlation, ρ is close to 1. If $\mathbf{x}$ and $\mathbf{y}$ have a strong negative linear correlation, ρ is close to -1. If there is no linear correlation or a weak linear correlation, ρ is close to 0. A value near zero means that there is a random, nonlinear relationship between the two vectors. In the CM tracking, we associate the eigenvectors at two different frequencies by comparing the magnitude of the correlation coefficients.

The third step in CM tracking is to associate the eigenvectors at two frequencies iteratively. An example is given to show this association procedure better. Assume we get a correlation coefficient matrix in the second step ($N = 4$):

$$\mathbf{C} = \begin{bmatrix} 0.0192 & 0.0152 & 0.0246 & 0.0235 \\ 0.2521 & 0.0135 & 0.4750 & 0.0278 \\ 0.0141 & 0.0175 & 0.0257 & 0.1060 \\ 0.0226 & 0.0026 & 0.0155 & 0.0165 \end{bmatrix} \qquad (2.122)$$

As can be observed, in the first and third columns, the maximum elements appear in the second row. Therefore, we can first assume that the first mode $\mathbf{J}_1(\omega_2)$ and the third mode $\mathbf{J}_3(\omega_2)$ at ω_2 have similar mode behavior with the second mode $\mathbf{J}_2(\omega_1)$ at ω_1. In the same way, we can observe that both $\mathbf{J}_2(\omega_2)$ and $\mathbf{J}_4(\omega_2)$ are associated with $\mathbf{J}_3(\omega_1)$. Therefore, the first round association judgment gives the tracked modes as shown in Figure 2.6.

However, we observe that the above mode association relationship is not unique for each mode. It does not clearly give the unique associated mode for mode pairs $\mathbf{J}_1(\omega_2)$ and $\mathbf{J}_3(\omega_2)$, and $\mathbf{J}_2(\omega_2)$ and $\mathbf{J}_4(\omega_2)$. Therefore, we need a second-round association judgment to compare the correlation strength of the above mode pairs.

From the correlation coefficient matrix in (2.122), we observe that $c_{23} > c_{21}$. It indicated that correlation strength between $\mathbf{J}_3(\omega_2)$ and $\mathbf{J}_2(\omega_1)$ is larger than that between $\mathbf{J}_1(\omega_2)$ and $\mathbf{J}_2(\omega_1)$. Therefore, $\mathbf{J}_3(\omega_2)$ is associated with $\mathbf{J}_2(\omega_1)$. In the same way, we can find that $c_{32} > c_{34}$. It indicates that $\mathbf{J}_2(\omega_2)$ is associated with $\mathbf{J}_3(\omega_1)$. Because $\mathbf{J}_2(\omega_2)$ and $\mathbf{J}_3(\omega_2)$ have been successfully associated to $\mathbf{J}_3(\omega_1)$ and $\mathbf{J}_2(\omega_1)$, we can get rid of the second and third rows and columns for further association judgment. As shown in Figure 2.7, we get a 2×2 matrix by removing these rows and columns.

Based on the reduced matrix in Figure 2.7, a third-round association judgment is required to associate the untracked modes. In the reduced matrix, $c_{14} > c_{44}$ indicates that $\mathbf{J}_4(\omega_2)$ is associated with $\mathbf{J}_1(\omega_1)$. Similarly, $c_{41} > c_{11}$ shows that $\mathbf{J}_1(\omega_2)$ should be associated with $\mathbf{J}_4(\omega_1)$. Therefore, we have successfully associated each of the

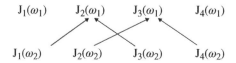

FIGURE 2.6 Association in the first round.

$$C = \begin{bmatrix} 0.0192 & 0.0152 & 0.0246 & 0.0235 \\ 0.2521 & 0.0135 & 0.4750 & 0.0278 \\ 0.0141 & 0.0175 & 0.0257 & 0.1060 \\ 0.0226 & 0.0026 & 0.0155 & 0.0165 \end{bmatrix}$$

$$C = \begin{bmatrix} 0.0192 & 0.0235 \\ 0.0226 & 0.0165 \end{bmatrix}$$

FIGURE 2.7 The reduced correlation coefficient matrix.

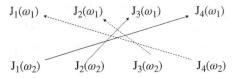

FIGURE 2.8 The association relationship between the eigenvectors at two different frequencies.

FIGURE 2.9 Configuration of a double-layered stacked microstrip antenna.

eigenvector at ω_2 to the eigenvectors at ω_1. In this example, the final association relationship for the eigenvectors at ω_1 and ω_2 is given in Figure 2.8.

In practical implementation of the CM analysis, there are usually much larger number of eigenvectors to track. The above procedure can be extended to CM analysis with more eigenvectors. The difference is that it may involve more iteration to achieve the final association relationship. A simple example is given in the following to show the differences between the tracked and untracked CMs.

2.8.2.2 A Modal Tracking Example
In order to show the application of CM tracking in the investigation of CMs across a wide frequency band, we consider a double-layered stacked microstrip antenna as shown in Figure 2.9. The bottom and top patches are printed on two layered dielectric slabs, respectively. This configuration is usually developed for achieving multiband or wideband antennas by exciting the resonant modes of the two patches simultaneously. Figure 2.10 shows the modal significances obtained from the CM analysis without and with the CM tracking

(a)

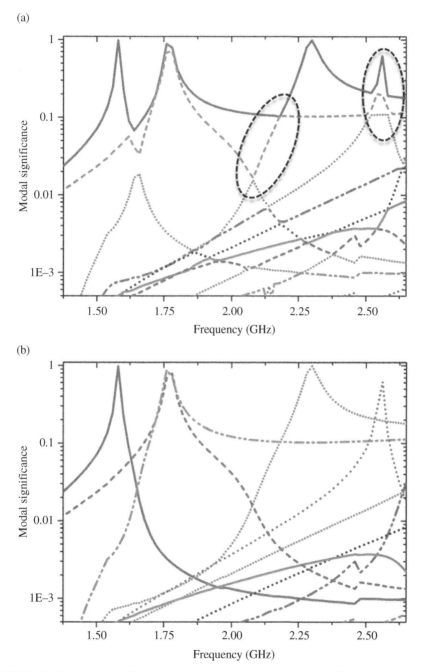

(b)

FIGURE 2.10   Modal significances with and without CM tracking. (a) CM analysis without CM tracking and (b) CM analysis with CM tracking.

(a)   (b)

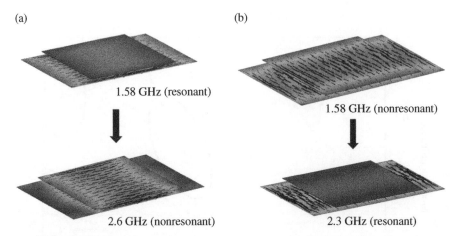

1.58 GHz (resonant)

1.58 GHz (nonresonant)

2.6 GHz (nonresonant)   2.3 GHz (resonant)

FIGURE 2.11   Characteristic currents of the (a) first and (b) fourth mode at the resonant and non-resonant frequencies.

procedure. Because the matrices for the generalized eigenvalue equation are only available at discrete frequencies, the modes at one frequency ω_1 may not be correctly associated with the modes at ω_2. As can be observed from Figure 2.10a, the lines that fall into the dotted circles are the same modes. However, they appear as different modes when modal significances are plotted across a frequency band. By applying the CM tracking, it can be observed from Figure 2.10b that all the modes vary smoothly across a frequency band. The modal significances belong to the same mode fall in the same line. It clearly shows that each mode evolves from non-resonance state to resonance state and finally arrives at the non-resonance state again.

The solid line in Figure 2.10b shows that the lowest mode resonates around 1.58 GHz, and the dotted line shows that the fourth mode resonates at around 2.3 GHz. In the case that the CM tracking procedure was not applied in the CM analysis, one will misunderstand from Figure 2.10a that all the dominant modes (modes with modal significance peaks) have the similar modal current distributions. As we shall see, the two resonant modes at 1.58 and 2.3 GHz are not the same mode. Figure 2.11 shows the characteristic currents of the two resonant modes around 1.58 and 2.3 GHz, respectively. As can be seen, the lowest mode at 1.58 GHz is the TM_{10} mode of the bottom patch, and the fourth mode at 2.3 GHz is the TM_{01} mode of the bottom patch. At non-resonant frequencies, the two modes have the similar current polarizations, but the top patch also contributes to the radiation and these modes exhibit as higher order modes.

2.9   NUMERICAL EXAMPLES

2.9.1   PEC Sphere

In general, closed form expressions of modal behavior of arbitrarily shaped PEC surfaces are not available. However, we are fortunate that the eigenvalues of a PEC sphere can be computed from closed formulations in the spherical mode theory [21]. It provides a straightforward investigation into the relationship between the CMs and the spherical mode theory. This section performs a comparison study to the CM analysis results and the analytical results from the spherical mode theory.

In the CM analysis, all the three CM formulations based on the EFIE, MFIE (AMFIE), and CFIE are applied to analyze a PEC sphere with 0.2 wavelength radius. The PEC sphere surface is discretized into 430 triangle elements, leading to 645 unknowns. Table 2.1 compares the eigenvalues solved from the CM formulations with the eigenvalues computed from exact formulations for the first eight CMs. First, we can observe that the eigenvalues solved from the EFIE-, MFIE-, and CFIE-based CM formulations are in good agreement. Therefore, one can use the three CM formulations independently for closed PEC surfaces. Second, we can also find from Table 2.1 that the eigenvalues computed from the CM theory agree well with the exact eigenvalues computed from the spherical mode theory. In this sense, the CM theory is a more general modal analysis approach for arbitrarily shaped PEC objects available. For conical PEC structures, the CMs are equivalent to the spherical modes. Third, we can observe that the combination coefficient α does not affect the accuracy of the eigenvalues a lot, ranging from very large MFIE contribution ($\alpha = 0.2$) to very large EFIE contribution ($\alpha = 0.8$). As expected, a careful investigation finds that the eigenvalues solved from CFIE with large α are closer to the EFIE results, and the eigenvalues solved from CFIE with small α are closer to the MFIE results. However, all of the CFIE CM analysis results have very small relative errors to the exact solution. Therefore, the CM analysis accuracy is not heavily dependent on the combination coefficient.

In addition to the very good agreement between the eigenvalues, the characteristic currents and characteristic fields (near fields and far fields) solved from the EFIE-, MFIE- (AMFIE), and CFIE-based CM formulations also achieve very good agreement. To keep concise, the following discussion will be based on the CMs solved from the EFIE-based CM formulation. One can also solve the same CMs from the MFIE- (AMFIE) and CFIE-based CM formulations.

Figure 2.12 shows the modal currents along with the associated modal fields for the first eight modes, that is, the TM_{01} mode, TE_{01} mode, TM_{12} mode, TM_{22} mode, TE_{02} mode, TE_{12} mode, TE_{22} mode, and TM_{03} mode. It shows that the TM_{01} mode and TE_{01} mode have the same current intensity distributions but orthogonal current polarizations. The TM_{01} mode has vertically polarized modal currents and the TE_{01} mode has horizontal polarized modal currents. Similarly, although the modal fields for the TM_{01} mode and TE_{01} mode have the same shape, they possess the orthogonal polarizations. The similar orthogonal properties are observed from the mode pairs TM_{12} and TE_{12}, TM_{22} and TE_{22}, respectively.

As the mode order increases from TM_{01}, TM_{12}, TM_{22} and finally to the highest calculated mode TM_{03}, there are more current nulls and peaks appearing in the characteristic currents. These current nulls and peaks produce radiation lobes in the modal fields. As observed from Figure 2.12p, there are totally six lobes in the modal fields of the TM_{03} mode. Because directional and omnidirectional radiation patterns are two of the most widely used radiation patterns in practical communication

TABLE 2.1 Eigenvalues of A PEC Sphere with 0.2 Wavelength Radius

Mode	Exact solution	EFIE-based CM	MFIE-based CM	CFIE-based CM: $\alpha = 0.2$	CFIE-based CM: $\alpha = 0.5$	CFIE based CM: $\alpha = 0.8$
TM_{01}	−1.082	−1.08361	−1.08334	−1.10317	−1.09471	−1.08739
TE_{01}	2.673	2.68668	2.67975	2.67495	2.64738	2.69225
TM_{11}	−11.00	−11.1549	−11.2515	−11.2813	−11.2023	−11.1432
TM_{12}	−11.00	−11.1261	−11.3374	−11.3845	−11.2557	−11.1705
TE_{02}	21.60	21.8687	22.1455	22.4787	21.9257	21.8592
TE_{12}	21.60	21.8536	22.199	22.1862	22.2523	21.6176
TE_{22}	21.60	21.8428	22.1605	22.097	22.8116	21.5433
TM_{03}	−284.40	−290.15	−297.372	−293.128	−291.434	−290.092

systems, higher-order modes with many server lobes usually need to be suppressed. This is one of the reasons that usually only lower-order modes are excited in antenna designs. Another reason is that lower-order modes always have small eigenvalues. It indicates that these lower-order modes will be more efficient if they are excited for radiation purpose.

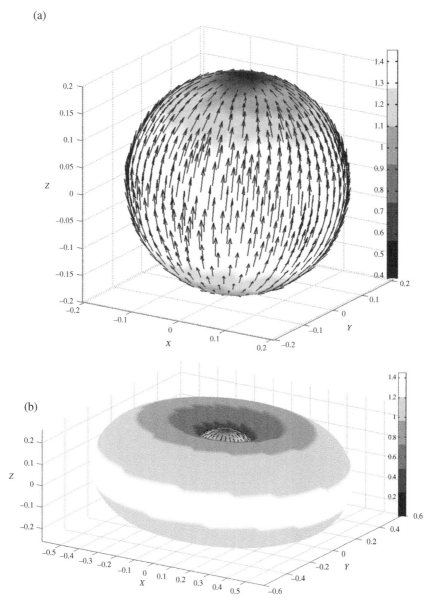

FIGURE 2.12 Characteristic modes of the PEC sphere with 0.2 wavelength radius. (a) Modal currents of the TM_{01} mode, (b) modal fields of the TM_{01} mode.

(c)

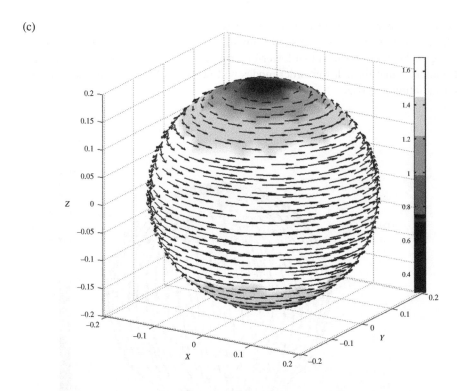

(d)

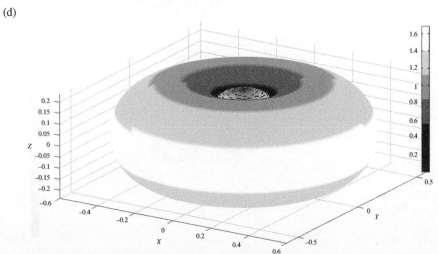

FIGURE 2.12   (*Continued*) (c) modal currents of the TE_{01} mode, (d) modal fields of the TE_{01} mode.

(e)

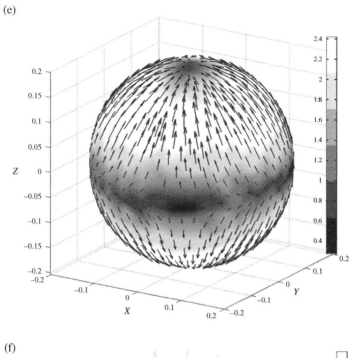

(f)

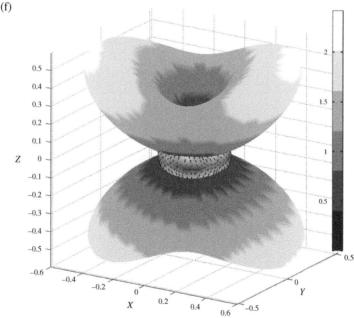

FIGURE 2.12 (*Continued*) (e) modal currents of the TM_{12} mode, (f) modal fields of the TM_{12} mode.

(g)

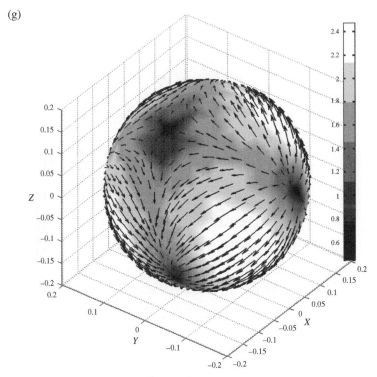

(h)

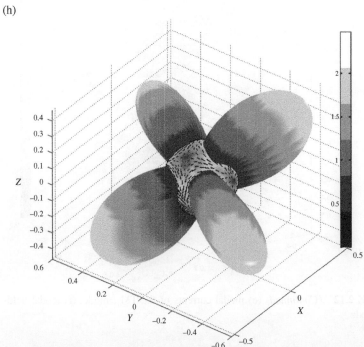

FIGURE 2.12 (*Continued*) (g) modal currents of the TM$_{22}$ mode, (h) modal fields of the TM$_{22}$ mode.

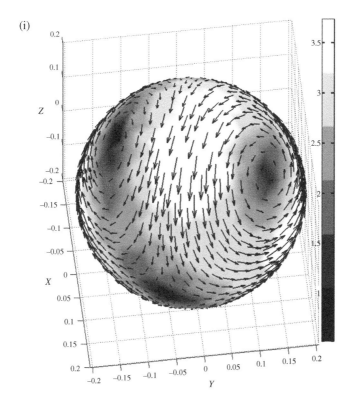

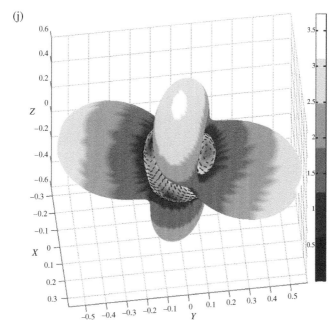

FIGURE 2.12 (*Continued*) (i) modal currents of the TE_{02} mode, (j) modal fields of the TE_{02} mode.

(k)

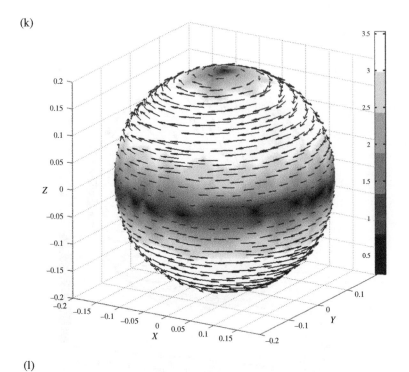

(l)

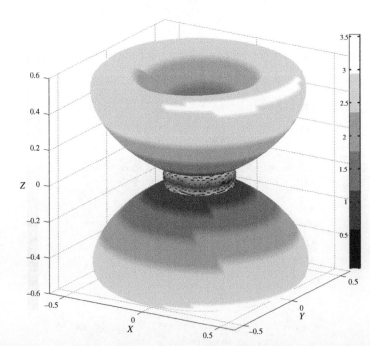

FIGURE 2.12   (*Continued*) (k) modal currents of the TE$_{12}$ mode, (l) modal fields of the TE$_{12}$ mode.

(m)

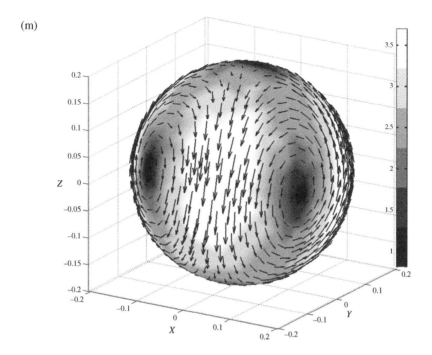

(n)

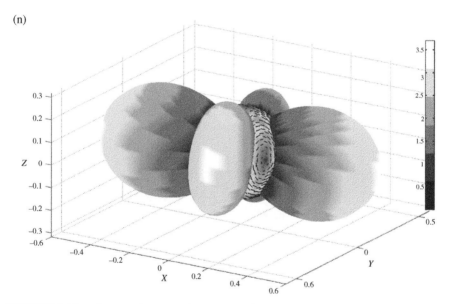

FIGURE 2.12 (*Continued*) (m) modal currents of the TE_{22} mode, (n) modal fields of the TE_{22} mode.

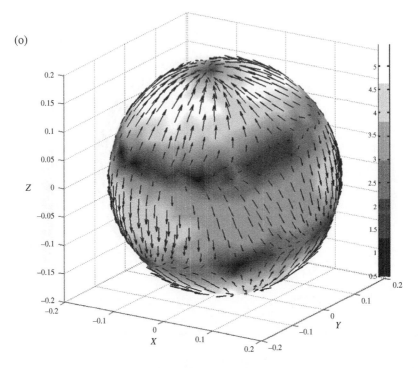

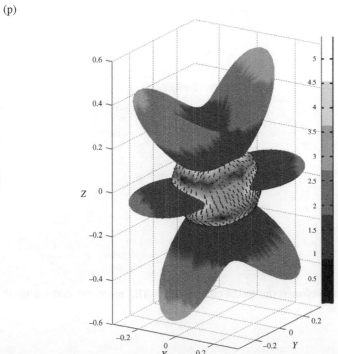

FIGURE 2.12 (*Continued*) (o) modal currents of the TM_{03} mode, and (p) modal fields of the TM_{03} mode.

2.9.2 Rectangular PEC Patch

In the preceding section, we have discussed the CM analysis for closed PEC sphere using the EFIE- and MFIE-based CM formulations. In this section, we will present the CM analysis for an open structure: a 100 mm × 40 mm rectangular plate. This rectangular plate is a typical ground plane for the chassis in mobile phones. In recent mobile phone antenna designs, the chassis is increasingly excited for antenna uses. The excitations of various modes of the chassis offer new design flexibilities. As compared to traditional monopole antennas in mobile phones, proper excitations for the ground plane's CMs greatly enhance the radiation efficiency. The CM analysis is carried out over the 820–4820 MHz frequency band. This frequency band covers many commonly used frequency band in civil communication systems, including the AMPS band (824–894 MHz), GSM band (880–960 MHz and 1850–1990 MHz), PCS band (1710–1880 MHz), and WCDMA band (1920–2170 MHz). Because the PEC plate is an open object, MFIE-based CM formulation is not suitable to this problem, and we will only employ the EFIE-based CM formulation in the analysis.

Figure 2.13 shows the eigenvalues across the frequency band. As can be seen, two resonances at 1300 and 2900 MHz are identified because of the zero eigenvalues. As the frequency increases, almost all the eigenvalues approach to zero. It reveals that there are more than one mode turning into the resonance states at higher frequencies. At these high frequencies, it is difficult to excite a single mode for good polarization purity. However, there is usually only one resonant mode in the low-frequency band.

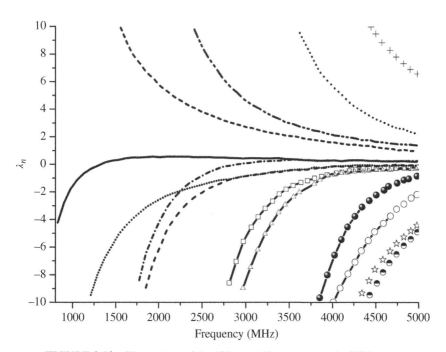

FIGURE 2.13 Eigenvalues of the 100 mm × 40 mm rectangular PEC plate.

In the lower frequency band, the lower-order modes are spaced far away with each other. They are easy to be excited for good polarization purity. This observation also explains the reason that, in most cases, only the first several low-order modes are excited for antenna designs.

Figure 2.14 shows the modal currents and modal fields at the first resonant frequency at 1300 MHz. As can be seen, the first mode is the TM_{10} mode and the second mode is TM_{01} mode. Both have broadside radiation patterns and have different current and field polarizations. These two modes are widely recognized in patch antenna designs and are usually excited for directive radiation patterns. The third and fourth modes are high-order modes. They have more complex polarization and radiation pattern properties and are usually suppressed in practical antenna designs.

Careful investigations of the modal currents and modal fields give useful information on how to feed the mobile phone chassis for designated radiation patterns. For the design of good MIMO antennas using a rectangular plate, two kinds of feeding designs were proposed to excite the different characteristic currents of the rectangular plate [27]. According to coupling mechanism of the electromagnetic power, the two feeding structures are known as the inductive coupling element (ICE) and capacitive coupling element (CCE), respectively. The ICE has to be placed at the location with maximum characteristic current for efficient mode excitation. Alternatively, CCE can be placed at the location with minimum characteristic current for efficient mode excitation. Figure 2.15 illustrates the two ways to excite the first mode as given in Figure 2.14a. In practical designs, ICE can be realized by cutting a slit at the current maxima and feeding the slit with a voltage source. As for the practical realization of the CCE, one can add a small element to the PEC plate and introduce a voltage source between the new introduced element and the PEC plate.

A careful comparison study showed that ICE offers advantages over CCE in terms of the mode purity [27]. The degraded polarization purity in the CCE scheme is mainly caused by the new element introduced to the original PEC plate. It will modify the original CMs of the PEC plate. However, there is no new element introduced in the ICE scheme. Therefore, only the original CM is excited, which keeps the mode purity of the CMs of the PEC plate.

2.9.3 Numerical Aspects of Mesh Density

In the MoM, the accuracy mainly depends on the mesh density, impedance matrix calculation, and solving of the resultant matrix equation. The CM analysis does not involve external source, and the accuracy is only dependent on the first two terms, that is, the mesh density and impedance matrix calculation. The accuracy of the impedance matrix calculation depends on the singularity treatment in the calculation of diagonal and near-diagonal elements. Readers can consult many CEM books [28–30] for more advanced topics on the singularity treatment in the MoM matrix element calculation. In this section, we only discuss the numerical aspects of the mesh density in the EFIE- and MFIE-based CM analysis. The investigations are based on both the closed and open PEC objects.

(a)

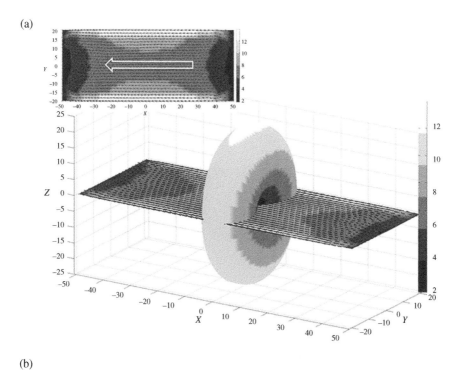

(b)

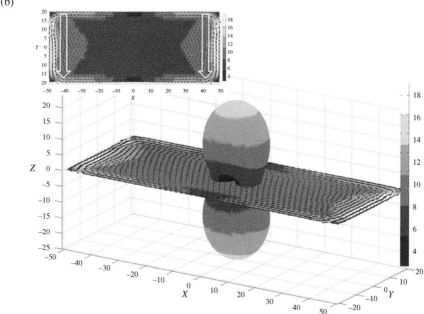

FIGURE 2.14 Characteristic modes of the 100 mm × 40 mm rectangular PEC plate at 1300 MHz. (a) $\mathbf{J}_1 \& \mathbf{E}_1$, (b) $\mathbf{J}_2 \& \mathbf{E}_2$.

(c)

(d)

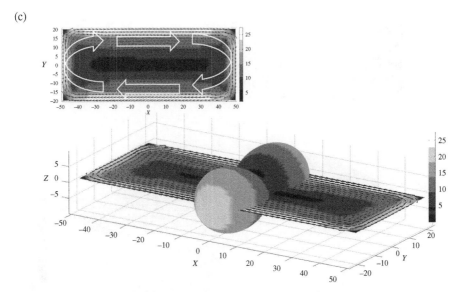

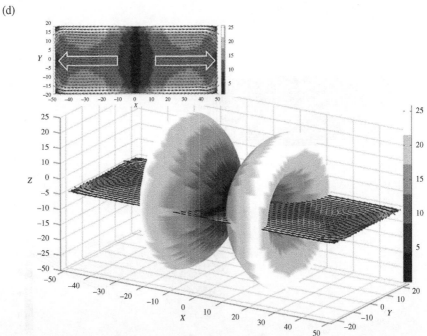

FIGURE 2.14 (*Continued*) (c) $\mathbf{J}_3 \& \mathbf{E}_3$, and (d) $\mathbf{J}_4 \& \mathbf{E}_4$.

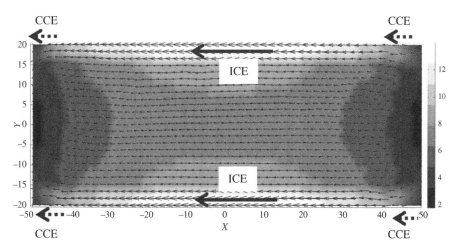

FIGURE 2.15 Inductive and capacitive coupling elements for the excitation of the first mode in Figure 2.14a.

TABLE 2.2   Different Mesh Densities for the 0.2 Wavelengths Radius PEC Sphere

Mesh density	No. of triangular elements	No. of MoM unknowns
$\lambda/8$	60	90
$\lambda/15$	276	414
$\lambda/30$	1486	2229

We first investigate the effect of the mesh density on the accuracy of the EFIE-based CM analysis for closed PEC objects. Three different mesh densities are used to mesh a PEC sphere with a radius of 0.2 wavelengths. Table 2.2 shows the number of triangle elements and MoM unknowns for the three different mesh densities. As the mesh density is increased, the number of RWG basis functions defined on the sphere surface varies from 90 to 2229. As mentioned in preceding sections, from the point view of numerical CM analysis, the number of CMs has to be the same as the number of MoM unknowns, although higher-order inefficient radiating modes are generally not computed. Evidently, different number of MoM unknowns in these mesh scheme will result in different set of CMs. However, the modal analysis for a specific structure should be a deterministic problem. This motivates us to see how the mesh density will affect the CM analysis results.

Figure 2.16 plots the eigenvalues of the first 40 modes for varying mesh density. It shows that the first 30 eigenvalues agree quite well among the three mesh densities. For much higher order modes, eigenvalues for the coarse mesh density of $\lambda/8$ deviate far away from those of the $\lambda/15$ and $\lambda/30$ cases. However, in the $\lambda/15$ and $\lambda/30$ cases, the eigenvalues for mode order higher than 30 also achieve satisfactory agreement with each other. It illustrates that high mesh density is required in CM analysis if very high-order modes are of interests. One should also note that $\lambda/8$ or $\lambda/10$ mesh density

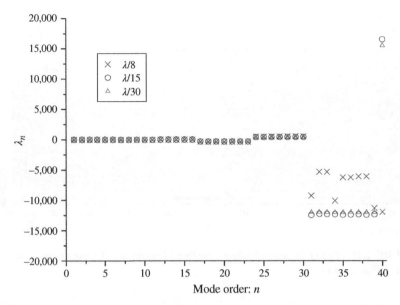

FIGURE 2.16 Eigenvalues of closed sphere surface for varying mesh density in EFIE-based CM formulation.

is usually enough to ensure reasonable accuracy in MoM simulations. The CM analysis, however, is developed to examine the resonant behavior of any interested modes (including higher order modes). Therefore, we suggest employing high mesh density in the CM analysis, especially in problems where the contribution of very high-order modes need to be examined seriously.

Similarly, we have also investigated the MFIE-based CM formulation. The CM analysis is also performed using the mesh shown in Table 2.2. Figure 2.17 illustrates the eigenvalues for the first 40 modes. As can be seen, the eigenvalues for the first 30 modes have very good agreement in all the mesh cases. As the mode order goes beyond 30, the eigenvalues obtained based on the $\lambda/8$ mesh density also deviate far away from those in the $\lambda/15$ and $\lambda/30$ cases. It is apparent that the eigenvalues in the $\lambda/8$ mesh density are not correct. We can also observe that, in the $\lambda/15$ and $\lambda/30$ cases, eigenvalues for mode order higher than 30 have not achieved satisfactory agreement. However, in the EFIE-based CM formulation, the two mesh densities result in good agreement in modes with order lower than 30. These investigations exhibit that the MFIE-based CM formulation is more sensitive to the mesh density. In terms of the stability, the EFIE-based CM formulation also outperforms the MFIE-based CM formulation. Together with the consideration that the EFIE-based CM formulation can be applied to both closed and open PEC surfaces, we suggest to implement the EFIE-based CM formulation in all the PEC problems for better accuracy.

Furthermore, the above CM analysis shows that although the numbers of unknowns in the three mesh cases are different, they give the same set of low-order

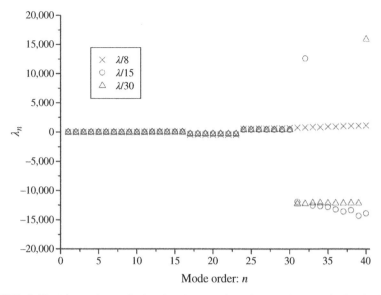

FIGURE 2.17 Eigenvalues of closed sphere surface for varying mesh density in the MFIE-based CM formulation.

TABLE 2.3 Different Mesh Densities for the 100 Mm × 40 Mm Rectangular PEC Plate

Mesh density	No. of triangular elements	No. MoM unknowns
$\lambda/8$	48	61
$\lambda/15$	180	249
$\lambda/30$	720	1038

CMs ($n < 30$). In most cases, we are only interested in the first several significant modes. As can be also observed from Figures 2.16 and 2.17, the numerically computed higher-order modes have very large eigenvalues. In the case that they are used to expand a given surface current through the modal solution formulation, their contributions are so weak that these modes can be ignored without significant errors. Therefore, if we do not care about very high-order modes, the mesh density is not a critical consideration in CM analysis.

As a further investigation, we consider the effect of the mesh density on the accuracy of EFIE-based CM analysis for open PEC objects. The investigation is based on the CM analysis for the 100 mm × 40 mm rectangular PEC plate at 3.0 GHz. Table 2.3 lists the mesh density we used in the investigation. The eigenvalues of the first 20 modes are shown in Figure 2.18. For the lower order modes ($n < 10$), all the mesh schemes give the same eigenvalues. However, the deviations among the three cases increase as the mode order gets higher. The observation also suggests fine mesh density has to be used to capture the behavior of high-order modes accurately. Moreover, as compared to the closed PEC sphere example, it observes that the CM analysis results for this open surface are more sensitive to the mesh density. To ensure

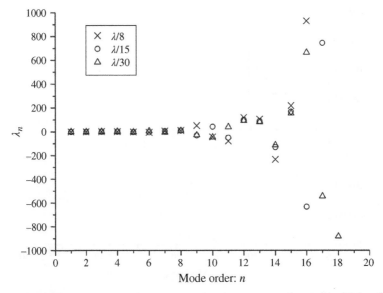

FIGURE 2.18   Eigenvalues of the PEC plate for varying mesh density.

the CM analysis accuracy of high-order modes, we suggest increasing the mesh density ($< \lambda / 15$) for open objects.

2.10   A FIRST GLANCE ON CM EXCITATIONS

As have been demonstrated in the CM analysis for PEC sphere and PEC rectangular plane, the characteristic currents and fields for each mode have their unique properties. The diversity of the CMs allows us to excite the desired mode for designated radiation patterns at certain frequency. We can excite a single mode for good polarization purity. We can also excite several modes at one frequency to obtain orthogonal far-field patterns. Furthermore, it is also possible to excite many modes at their own resonant frequencies for multiband or even wideband antennas designs. This section will discuss the CM excitation of a planar inverted-F structure for multiband antenna designs.

Planar inverted-F antenna, which is also known as PIFA [31], is extremely popular in mobile and wireless communications. It features important properties including low profile, low cost. Most importantly, it has very wide impedance BW. Figure 2.19 shows the geometry of a multiband planar inverted-F antenna. It is developed from a finite ground plane. Parasitic radiators that are coupled to the driven element and shorting post are the extensively used techniques to extend the impedance BW. A voltage source can be placed between the feed plate and the ground plane to excite the structure.

Figure 2.20 shows the simulated VSWR against the frequency over 1.5–6 GHz using HFSS, a commercial electromagnetic simulation software based on the FEM.

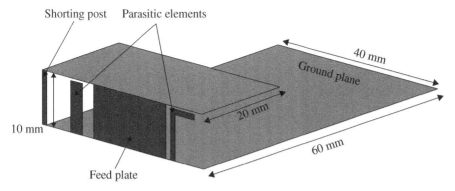

FIGURE 2.19   Geometry of a multiband planar inverted-F antenna.

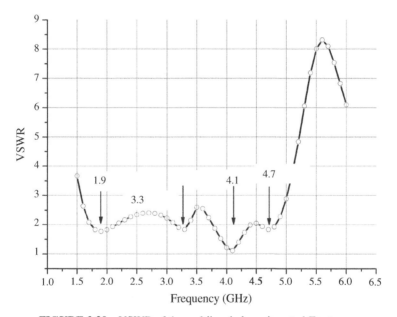

FIGURE 2.20   VSWR of the multiband planar inverted-F antenna.

It is observed that there are four resonances at 1.9, 3.3, 4.1, and 4.7 GHz, respectively. Figure 2.21 shows the current distribution and radiation patterns at these resonant frequencies. As can be seen, the four resonances are contributed by different parts of the planar inverted-F structure:

1. The resonant currents at 1.9 GHz mainly distribute on the shorting post and the right edge of the feed plate;

2. The resonant currents at 3.3 GHz mainly distribute on the inverted-F parasitic element and the right edge of the feed plate;

(a)

(b)

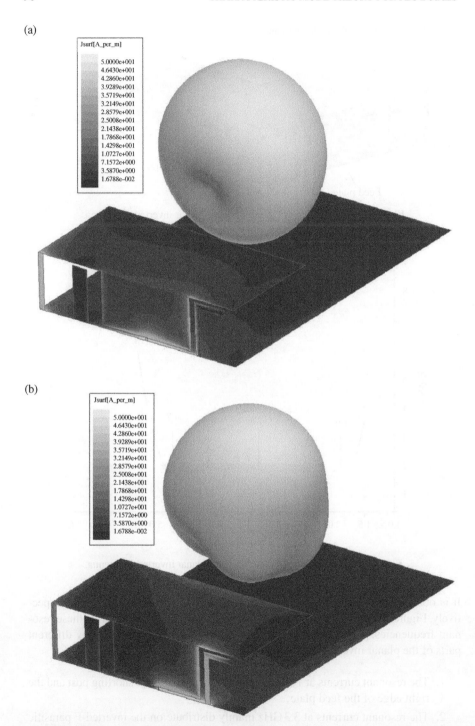

FIGURE 2.21 Current distributions and radiation patterns at resonant frequencies. (a) 1.9 GHz (b) 3.3 GHz.

(c)

(d)

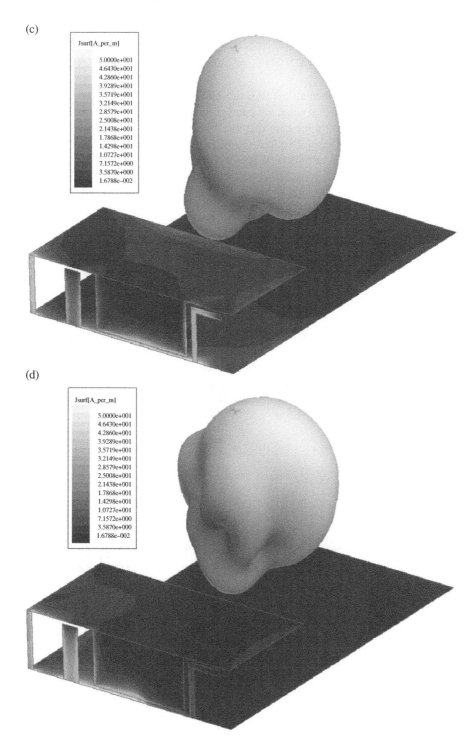

FIGURE 2.21 (*Continued*) (c) 4.1 GHz, and (d) 4.7 GHz.

3. The resonant currents at 4.1 GHz mainly distribute on the inverted-F parasitic element;

4. The resonant currents at 4.7 GHz mainly distribute on the straight parasitic strip.

As can be observed, when all the four resonances are excited simultaneously by the properly placed voltage source, we can get the current distribution within each operating band. However, in the case that the voltage source is not properly placed, full-wave simulation will not give the resonant currents and resonant frequencies. Indeed, it is difficult for an inexperienced antenna designer to excite an antenna with a properly designed feeding structure for multiband operation. There are many parameter studies in the literature to investigate the physics behind the wide impedance BW. In these studies, both the feeding location and the dimensions of the PIFA are investigated. However, it is still impossible to get the general design guidelines for PIFA designs. In this section, we will look at PIFA's wide bandwidth from the point of view of CMs. It is expected that the CMs will give more explanations on the wideband performance of the PIFA. As we shall see, the eigenvalues reveal the resonant frequencies of the structure, and the characteristic currents clearly indicate how to place the source to excite all of the desired CMs.

Figure 2.22 shows the eigenvalues of the PIFA over 1.5–6 GHz. Following the physical interpretations for the zero eigenvalues, we observe that the resonant frequencies below 4.25 GHz locate at 1.6, 3.0, and 4.1 GHz, respectively. These resonant frequencies are determined by the PIFA structure itself, and they are independent of any external sources. As compared to the preceding full-wave simulation results, the CM analysis results achieve reasonable agreement with them. As can be seen, the

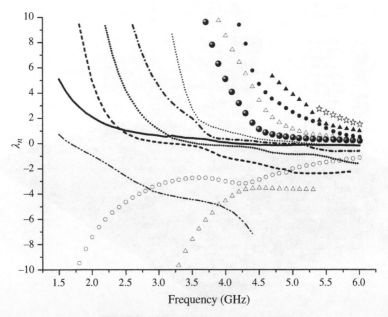

FIGURE 2.22 Eigenvalues of the PIFA.

CM theory provides a new approach to predict the natural resonant frequencies without any need for external sources. When the frequency increases beyond 4.25 GHz, there are many higher-order modes resonant over the entire high-frequency band. It leads to difficulties in the excitation of one single mode over 4.25 GHz. When an external source is placed on the PIFA, all these modes will be excited. Because of these continuously spaced higher-order modes, the radiation pattern at 4.7 GHz in Figure 2.21d suffers some pattern distortion.

Figure 2.23 presents the characteristic currents of the dominant modes at each resonant frequency. The characteristic currents clearly show the part of the PIFA

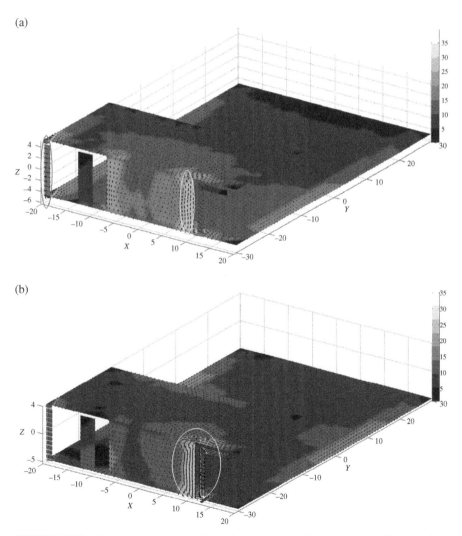

FIGURE 2.23 Dominant characteristic modes at resonant frequencies. (a) Characteristic currents at 1.9 GHz, (b) characteristic currents at 3.3 GHz.

(c)

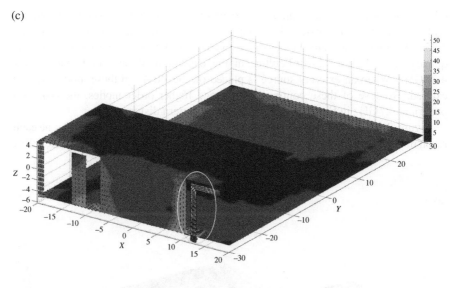

(d)

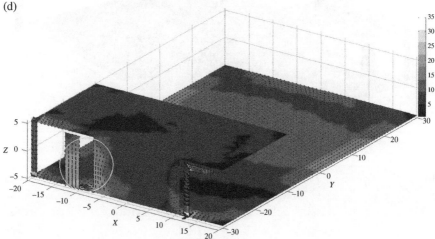

FIGURE 2.23 (*Continued*) (c) characteristic currents at 4.1 GHz, and (d) characteristic currents at 4.7 GHz.

structure that contributes to the resonant currents at each resonant frequency. It shows the following:

- The dominant characteristic currents at 1.9 GHz mainly distribute on the shorting post and the right edge of the feed plate;
- The dominant characteristic currents at 3.3 GHz mainly distribute on the inverted-F parasitic element and the right edge of the feed plate;

- The dominant characteristic currents at 4.1 GHz mainly distribute on the inverted-F parasitic element;
- The dominant characteristic currents at 4.7 GHz mainly distribute on the straight parasitic strip.

As compared to the full-wave simulated currents, the characteristic currents show the same results in the investigation of the main contributor for each resonance. Although the excitation structure is not considered, we can accurately predict the resonant currents through the CM analysis. Moreover, the characteristic currents illustrate how to feed the PIFA structure for multi/wide band operation. Generally, we have to place voltage source on the location with strong characteristic currents to excite the desired mode. However, the strong currents for different modes appear on different parts of the PIFA. A possible way might be to place the source on the wide feed plate and expect the parasitic elements and shorting post will couple energy from the wide feed plate. Through this excitation scheme, the dominant modes in the frequency band of 1.9, 3.3, 4.1, and 4.7 GHz are excited. In short, the PIFA design is a typical multiband antenna design by using different dominant modes at many different frequency bands.

2.11 SUMMARY

Starting with the electrical field integral equation (EFIE) and MFIE, this chapter formulates two generalized eigenvalue equations for the CM analysis of PEC objects. Important properties of CMs and physical interpretations to CM quantities are followed. Key algorithms for the solving of generalized eigenvalue equation and modal tracking are addressed for computer code implementations. Numerical examples are presented to examine the EFIE- and MFIE-based CM analysis results. Specifically, the relationship between the mesh density and CM analysis accuracy is investigated to give general guidelines on choosing of mesh density. Illustrative explanations on the eigenvalues, modal currents, and modal fields are presented to show valuable information in the CM analysis results. Comparison study is also conducted based on the full wave simulations with external source. It shows that the CM theory provides a source-free approach to investigating the resonant behavior and radiation performance of an electromagnetic structure. These numerical examples give a first glance on how the CM theory can be applied in antenna engineering.

REFERENCES

[1] R. F. Harrington, *Field Computation by Moment Methods*. New York, NY: Macmillan, 1968.

[2] K. S. Yee, "Numerical solution of initial boundary value problems involving Maxwell's equations in isotropic media," *IEEE Trans. Antennas Propagat.*, vol. 14, no. 3, pp. 302–307, May 1966.

[3] K. S. Kunz and R. J. Luebbers, *The Finite Difference Time Domain Method for Electromagnetics*. Boca Raton, FL: CRC Press, 1994.

[4] A. Taflove and S. C. Hagness, *Computational Electrodynamics: The Finite Difference Time Domain Method* (3rd edition). Norwood, MA: Artech House, 2005.

[5] P. P. Silvester and R. L. Ferrari, *Finite Elements for Electrical Engineers* (3rd edition). Cambridge, UK: Cambridge University Press, 1996.

[6] J.-M. Jin, *The Finite Element Method in Electromagnetics* (2nd edition). New York, NY: John Wiley & Sons, Inc., 2002.

[7] R. J. Garbacz, "Modal expansions for resonance scattering phenomena," *Proc. IEEE*, vol. 53, pp. 856–864, 1965.

[8] R. F. Harrington and J. R. Mautz, "Theory of characteristic modes for conducting bodies," *IEEE Trans. Antennas Propagat.*, vol. AP–19, no. 5, pp. 622–628, Sep. 1971.

[9] J. V. Bladel, *Electromagnetic Fields* (2nd edition). Hoboken, NJ: Wiley-IEEE Press, May 2007.

[10] R. F. Harrington, "Origin and development of the method of moments for field computation," in *Computational Electromagnetics: Frequency-Domain Method of Moments* (E. K. Miller, L. Medgyesi-Mitschang, and E. H. Newman, eds), New York, NY: IEEE Press, pp. 43–47, 1992.

[11] ANSYS Corporation. Homepage of ANSYS Corporation. Available at www.ansys.com. Accessed January 23, 2015.

[12] MSC Software Corporation. Homepage of MSC Software Corporation. Available at http://www.mscsoftware.com/product/patran. Accessed January 23, 2015.

[13] S. M. Rao, D. R. Wilton, and A. W. Glisson, "Electromagnetic scattering by surfaces of arbitrary shape," *IEEE Trans. Antennas Propagat.*, vol. 30, no. 3, pp. 409–418, May 1982.

[14] J. M. Jin, *Theory and Computation of Electromagnetic Fields*. Hoboken, NJ: Wiley-IEEE Press, 2010.

[15] J. Ethier, E. Lanoue, and D. McNamara, "MIMO handheld antenna design approach using characteristic mode concepts," *Microw. Opt. Tech. Lett.*, vol. 50, pp. 1724–1727, 2008.

[16] E. H. Newman, "Small antenna location synthesis using characteristic modes," *IEEE Trans. Antennas Propag.*, vol. AP-21, no. 4, pp. 530–531, Jul. 1979.

[17] A. H. Nalbantoglu, "New computation method for characteristic modes," *Electron. Lett.*, vol. 18, no. 23, pp. 994–996, Nov. 1982.

[18] L. M. Correia and A. M. Barbosa, "An approximate (symmetric) magnetic field integral equation for the analysis of scattering by conducting bodies," in IEEE International Symposium Antennas and Propagation Society, Dallas, TX, vol. 2, pp. 902–905, May 7–11, 1990.

[19] L. M. Correia, "A comparison of integral equations with unique solution in the resonance region for scattering by conducting bodies," *IEEE Trans. Antennas Propag.*, vol. 41, no. 1, pp. 52–58, Jan. 1993.

[20] A. D. Yaghjian, "Augmented electric and magnetic field equations," *Radio Sci.*, vol. 16, no. 6, pp. 987–1001, Nov.–Dec. 1981.

[21] R. F. Harrington and J. R. Mautz, "Computation of characteristic modes for conducting bodies," *IEEE Trans. Antennas Propagat.*, vol. AP-19, no. 5, pp. 629–639, Sep. 1971.

[22] ARPACK software. Homepage of ARPACK. Available at http://www.caam.rice.edu/software/ARPACK/. Accessed January 23, 2015.

[23] R. B. Lehoucq, D. C. Sorensen, and C. Yang, *ARPACK Users Guide: Solution of Large-Scale Eigenvalue Problems with Implicitly Restarted Arnoldi Methods*. Philadelphia, PA: SIAM, 1998.

[24] LAPACK. Homepage of LAPACK. Available at http://www.netlib.org/lapack/. Accessed January 23, 2015.

[25] M. Capek, P. Hazdra, P. Hamouz, and J. Eichler, "A method for tracking characteristic numbers and vectors," *Prog Electromagn. Res. B*, vol. 33, pp. 115–134, 2011.

[26] B. D. Raines and R. G. Rojas, "Wideband characteristic mode tracking," *IEEE Trans. Antennas Propagat.*, vol. 60, no. 7, pp. 3537–3541, Jul. 2012.

[27] R. Martens, E. Safin, and D. Manteuffel, "Inductive and capacitive excitation of the characteristic modes of small terminals", in Antennas and Propagation Conference (LAPC), 2011 Loughborough, UK, pp. 1–4, Nov. 2011.

[28] X.-Q. Sheng and W. Song, *Essentials of Computational Electromagnetics*. Singapore: John Wiley & Sons Singapore Pte. Ltd, 2012.

[29] B. H. Jung, T. K. Sarkar, Y. Zhang, et al., *Time and Frequency Domain Solutions of* EM Problems Using Integral Equations and a Hybrid Methodology. Hoboken, NJ: Wiley-IEEE Press, 2010.

[30] W. C. Gibson, *The Method of Moments in Electromagnetics* (2nd edition). Boca Raton, FL: Chapman and Hall/CRC, 2014.

[31] T. Taga and K. Tsunekawa, "Performance Analysis of a Built-In Planar Inverted F Antenna for 800 MHz Band Portable Radio Units," *IEEE Journal on Selected Areas in Communications*, vol. 5, no. 5, pp. 921–929, Jun. 1987.

3

CHARACTERISTIC MODE THEORY FOR ANTENNAS IN MULTILAYERED MEDIUM

3.1 BACKGROUNDS

In Chapter 2, we have discussed the CM theory and its fundamental applications for perfectly electrically conducting (PEC) objects in free space. In this chapter, we will discuss the characteristic mode (CM) theory for another type of electromagnetic problems: microstrip patch antennas with multilayered medium. We will investigate and develop many microstrip patch antennas from the point view of CMs.

In 1953, Deschamps and Sichak [1] proposed the concept of microstrip patch antenna. Long time after that, microstrip patch antennas underwent extensive research and tremendous new developments were made since the 1970s. In recent times, microstrip patch antennas have found numerous applications in a variety of wireless communication systems, due to their attractive features like low profile, low cost, easy to conformal to shaped surface, and easy to integrate with circuit elements. A large number of papers have been published on the topic of microstrip patch antennas. Many books are also available for microstrip patch antennas [2–8]. Readers are encouraged to read these books for technical details on broadband and compact microstrip antenna designs.

Figure 3.1 shows the basic configuration of a microstrip patch antenna. It consists of a radiating patch supported by a grounded dielectric substrate. In principle, the patch can be in arbitrary shapes. In practical designs, the most popular patches used in microstrip patch antennas include the rectangular patch, the circular patch, and

Characteristic Modes: Theory and Applications in Antenna Engineering, First Edition.
Yikai Chen and Chao-Fu Wang.
© 2015 John Wiley & Sons, Inc. Published 2015 by John Wiley & Sons, Inc.

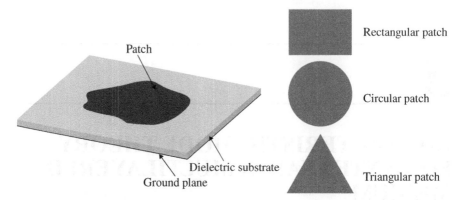

FIGURE 3.1   Basic configuration of a microstrip patch antenna.

the triangular patch. By carefully designing the feedings, either linearly or circularly polarized (CP) waves can be excited using these patches for radiation purpose. We will investigate the resonant behavior and radiation performance of these basic patch antennas through the CM analysis.

Figure 3.2 displays four typical feeding structures for microstrip patch antennas. There are different ways to transfer RF power from the transmission system to the antenna. In Figure 3.2a, the coaxial probe feed comprises a probe extended from the inner conductor of a coaxial cable. This probe is directly connected to the patch. The outer conductor of the coaxial cable is soldered to the ground. Therefore, in the case of very thin dielectric substrate, the coaxial probe feed provides a voltage source between the patch and the ground plane. Impedance matching performance can be tuned by adjusting the location of the feed point.

Figure 3.2b shows the microstrip line feed for microstrip patch antennas. In this feed method, the microstrip line directly connects to the patch to offer a voltage source at the connection point. As compared to the coaxial probe feed, it avoids the solder point and is very convenient for the integration of antennas with other circuit element on the same surface.

Figure 3.2c presents the proximity coupled feed for a microstrip patch antenna. Different from the microstrip line feed as shown in Figure 3.2b, the patch and the microstrip line in the proximity coupled feed are printed on two separate substrates. In other words, the patch and the microstrip line are not on the same plane. There is no physical connection between the microstrip line and the patch. Owing to the close proximity between the patch and the microstrip line, electromagnetic energy is coupled from the microstrip line to the patch. As compared to the first two feed structures, the proximity coupled feed may produce wider impedance bandwidth due to the increased height between the patch and the ground plane. It also avoids the solder point and is easy for the integrations with circuit elements.

The aperture coupled feed shown in Figure 3.2d is another attractive feed structure to achieve broad impedance bandwidth in microstrip patch antennas. The ground

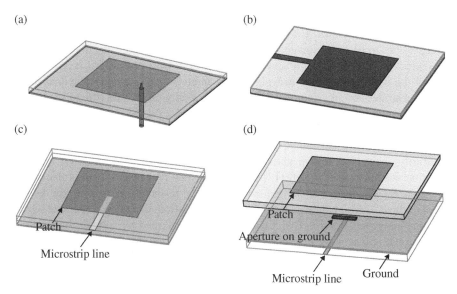

FIGURE 3.2 Typical feeding structures for microstrip patch antennas. (a) Coaxial probe feed, (b) microstrip line feed, (c) proximity coupled feed, and (d) aperture coupled feed.

plane separates the two substrates. It serves as the common ground for the radiating patch and the microstrip feed line. The radiating patch resides on the top of the upper substrate, whereas the microstrip feed line is printed on the bottom of the lower substrate. The electromagnetic energy is coupled from the microstrip feed line to the radiating patch through the aperture coupling. The ground plane isolates the microstrip line from the radiating patch. It reduces the spurious feed line radiation and ensures the polarization purity. The aperture on the ground plane can be also designed as an additional radiator. In this case, the aperture facilitates the dual-band antenna designs and broadband antenna designs, depending on whether the resonant frequency of the slot is far away from or close to the resonant frequency of the radiating patch.

As can be seen, antenna engineers have to properly choose the dielectric material, the substrate thickness, the shape, and size of the patch, and the feeding structure to meet specific antenna performances. Due to the complicated electromagnetic interactions among the patch, dielectric substrate, and the feeding structure, there is generally no exact formulation to determine these design parameters.

Over the past decades, a number of methods have been developed to determine the resonant frequencies and/or far-field radiation performance of microstrip patch antennas [9–12]. These methods can be roughly categorized into the following two types: the analytical solution and the full-wave analysis methods. Analytical solutions to microstrip patch antennas always simplify the real antenna model and are developed based on many assumptions. Cavity model is a representative analytical method [9]. It provides good intuitive explanations to the operations of canonic microstrip patch antennas. The basic assumption in cavity model is that the substrate thickness has to be much smaller than the wavelength. It ensures that the field inside the dielectric material has only a vertical component. Unfortunately, it is not easy to amend the cavity model for irregular patch antennas or microstrip antennas with thick substrate. As an example, the corrected cavity model for irregular-shaped microstrip antennas involve very complicated formulas [9, 13].

Full-wave analysis methods solve the Maxwell's equations with particular boundary conditions and sources. In principle, they can provide accurate results for arbitrarily shaped microstrip antennas. For example, a method based on integral equation was proposed in Ref. [12] to determine the resonances of irregular patch antennas. By sweeping a wide frequency band, the resonant frequencies are recognized as those at which the real part of the surface-current density is maximum while the imaginary part is zero. The current density is obtained by solving the integral equation with plane wave incidence. However, the current distributions for resonant modes are rather confusing and in some cases would be misleading because they depend on the incidence angle and polarization of the incident fields. Such full-wave analysis methods also provide little physical insights into the radiation mechanism of microstrip patch antennas.

The CM theory was initially developed to investigate the intrinsic resonant behavior of PEC objects in free space [14]. Without any exceptions, extending the CM theory to antennas in multilayered medium will provide a general approach to the analysis of microstrip patch antennas. In recent years, people have attempted to extend the conventional CM theory [14] to the analysis of microstrip patch antennas [15–17]. However, these early attempts have some limitations. In Refs. [15, 16], resonant frequency of the fundamental TM_{10} mode for rectangular patch antennas with air substrate was computed using the conventional CM theory. However, this approach was not able to give the resonant frequencies for high-order modes. In Ref. [17], a preliminary study for rectangular patch antennas with air substrate was reported. It was pointed out that modes with closed loop current distributions are non-radiating modes. However, these non-radiating modes also have eigenvalues close to zero. It may lead to a misleading conclusion that these mode with zero eigenvalues can also radiate efficiently. It also contradicts with the definition and physical interpretations of the eigenvalues in the CM theory.

In this chapter, we propose a new CM theory for the calculation of resonant frequencies and radiating modes of irregularly shaped patch antennas in multilayered medium. The generalized eigenvalue equation is developed from the mixed-potential integral equation (MPIE) with the spatial domain Green's functions (GFs) of multilayered medium. The GF takes into account the multilayered medium environment of the patches. Therefore, the CM theory based on this Green's function can accurately characterize the resonant frequencies, modal currents, and modal far fields of both fundamental and high-order radiating modes. It also avoids confusing non-radiating modes as computed from the conventional CM theory. For analysis purpose, the developed CM theory can well explore the underlying physics of circularly polarized antennas, multi-mode antennas, multi-band antennas, and stacked antennas. In the design sense, this CM theory is helpful for locating the feeding locations and optimizing the shape and the size of radiating patch for the enhancement of various radiation performances.

3.2 CM FORMULATION FOR PEC STRUCTURES IN MULTILAYERED MEDIUM

The conventional CM theory was initially developed for PEC objects in free space [14]. The resultant CMs define a set of orthogonal currents that can be used to expand the induced currents due to external sources. As opposed to the electrical field integral equation (EFIE) for PEC objects in free space, the MPIE with spatial domain GFs of multilayered medium offers the capability of accurately modeling the surface currents on metallic conductors buried within the multilayered medium [18–22]. The ground plane and dielectric substrates in the MPIE are assumed to be infinite in the transverse directions. Thus it is suitable for the analysis of microstrip patch antennas printed on dielectric substrate material with large transverse sizes. The major advantage of the MPIE is its lower memory requirement. The basis functions are only defined on the radiating patches.

Figure 3.3 shows the geometry of an irregularly shaped patch antenna. Without loss of generality, we assume the radiating patch resides on the interface of two layered dielectric substrates. The ground plane and the dielectric slabs are assumed to be infinite in the transverse directions. Considering the boundary condition on the PEC, that is, the tangential components of the electric field on the PEC surfaces equal to zero, the MPIE that relates the unknown current $\mathbf{J}$ and incident field $\mathbf{E}^i$ can be written as follows [18–22]:

$$\mathbf{E}^i_{\text{tan}} = \left[j\omega\mu_0 \left\langle G_\text{a}, \mathbf{J} \right\rangle - \frac{1}{j\omega\varepsilon_0} \nabla \left\langle G_\text{q}, \nabla' \cdot \mathbf{J} \right\rangle \right]_{\text{tan}} \tag{3.1}$$

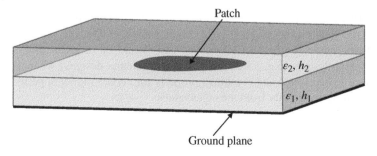

Patch

ε_2, h_2

ε_1, h_1

Ground plane

FIGURE 3.3   Geometry of an irregularly shaped patch antenna.

where G_a and G_q are the spatial domain GFs for the vector and scalar potentials, respectively. The closed form expression of the spatial domain GFs can be obtained using the discrete complex image method [18–20].

By applying the standard Galerkin's procedure to the MPIE in Equation (3.1), the elements of the resultant impedance matrix $\mathbf{Z}$ can be expressed as follows:

$$\mathbf{Z}_{mn} = j\omega\mu_0 \iint_{Tm} \iint_{Tn} \mathbf{f}_m(\mathbf{r}) \cdot \mathbf{f}_n(\mathbf{r}') G_a(\mathbf{r},\mathbf{r}')\,d\mathbf{r}'d\mathbf{r}$$
$$+ \frac{1}{j\omega\varepsilon_0} \iint_{Tm} \iint_{Tn} \nabla \cdot \mathbf{f}_m(\mathbf{r}) \nabla' \cdot \mathbf{f}_n(\mathbf{r}') G_q(\mathbf{r},\mathbf{r}')\,d\mathbf{r}'d\mathbf{r} \tag{3.2}$$

where $\mathbf{f}_m(\mathbf{r})$ is the well-known RWG basis functions defined over triangle pairs on the radiating patch. Following the procedure of the conventional CM theory for PEC objects [14], a generalized eigenvalue equation for the MPIE can be written as:

$$\mathbf{X}\mathbf{J}_n = \lambda_n \mathbf{R}\mathbf{J}_n \tag{3.3}$$

where λ_n is the eigenvalue associated with each characteristic current $\mathbf{J}_n$. $\mathbf{R}$ and $\mathbf{X}$ are the real and imaginary Hermitian parts of the matrix $\mathbf{Z}$, respectively:

$$\mathbf{R} = \frac{\mathbf{Z} + \mathbf{Z}^*}{2} \tag{3.4}$$

$$\mathbf{X} = \frac{\mathbf{Z} - \mathbf{Z}^*}{2j} \tag{3.5}$$

Similar to the eigenvalues for PEC objects in free space, modes with small $|\lambda_n|$ are effective radiating modes of the patch antennas, while those with large $|\lambda_n|$ are poor radiating modes to excite for radiation purpose. Other than that the eigenvalues tell how the mode radiates; modal significance is also a helpful indicator and can be defined following the case of PEC objects in free space:

$$\mathrm{MS}_n = \frac{1}{|1 + j\lambda_n|} \tag{3.6}$$

Similar to the PEC case, modal significance represents the contribution of a particular mode to the total radiation when an external source is applied. In addition, the characteristic angle can be also introduced and should be of great importance to many antenna designs (e.g., circularly polarized antenna designs) where multiple modes need to be excited and combined properly for certain radiation performance. It can be computed directly from the eigenvalues:

$$\alpha_n = 180° - \tan^{-1} \lambda_n \tag{3.7}$$

From a physical point of view, the characteristic angle models the phase angle between the real characteristic current and its associated characteristic fields.

The orthogonality properties of the CMs are also similar to those in the conventional CM theory [14]. Specifically, the characteristic currents are orthogonal with each other over the radiating patch, and the characteristic fields are orthogonal with each other over the radiation sphere in the infinity. Meanwhile, modal solutions in terms of the characteristic currents and characteristic fields are also kept in the CM theory for antennas with multilayered medium. They are useful to expand the radiating currents and far fields to examine the contributions from each mode.

3.3   RELATIONSHIP BETWEEN CAVITY MODEL AND CHARACTERISTIC MODES

The rectangular patch is perhaps the most commonly used patch shape in the area of microstrip antennas. Cavity model is one of the most convenient approaches to analyze the resonant behavior of rectangular microstrip antennas [23, 24]. In order to investigate the relationship between the cavity model and the characteristic modes, this section presents a comparison study to the cavity model and the characteristic modes through a rectangular microstrip patch antenna as shown in Figure 3.4. Full-wave simulation results will also be given to show that the dominant CMs at each resonant frequency can be excited through a coaxial probe feed.

As shown in Figure 3.4, the rectangular patch has the following dimensions: $L=76\,\text{mm}$ and $W=50\,\text{mm}$. It is printed on a single-layered dielectric substrate with a relative permittivity of $\varepsilon_r=3.38$ and thickness $h=1.524\,\text{mm}$. In the cavity model, resonant frequencies f_{mn} of TM_{mn} modes for a rectangular patch antenna can be determined from:

$$f_{mn} = \frac{c}{2\sqrt{\varepsilon_{\text{eff}}}} \sqrt{\left[\frac{m}{L+2\Delta L}\right]^2 + \left[\frac{n}{W+2\Delta W}\right]^2} \tag{3.8}$$

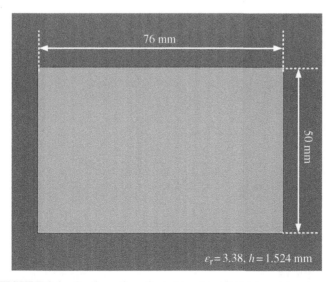

FIGURE 3.4   Configuration of a rectangular microstrip patch antenna.

where ΔL and ΔW are the equivalent length and width account for the fringing fields at the open ends along the patch length and width, respectively. ε_{eff} is the effective dielectric constant and can be calculated from:

$$\varepsilon_{\text{eff}} = \frac{\varepsilon_{\text{r}}+1}{2} + \frac{\varepsilon_{\text{r}}-1}{2}\left(1+10\frac{h}{W}\right)^{-\alpha\beta} \tag{3.9}$$

where α and β are the empirical parameters. They are given by:

$$\alpha = 1 + \frac{1}{49}\log_{10}\frac{(W/h)^4+(W/52h)^2}{(W/h)^4+0.432} + \frac{1}{18.7}\log_{10}\left[1+(W/18.1h)^3\right] \tag{3.10}$$

$$\beta = 0.564\left(\frac{\varepsilon_{\text{r}}-0.9}{\varepsilon_{\text{r}}+3}\right)^{0.053} \tag{3.11}$$

Following Equation (3.8), the resonant frequencies of the fundamental TM_{10} mode and high-order modes TM_{01}, TM_{11}, and TM_{20} of the rectangular patch antenna locate at 1.075, 1.605, 1.955, and 2.145 GHz, respectively. In order to excite these modes simultaneously, a coaxial probe feed is placed at a properly chosen feed point. Figure 3.5 shows the resultant reflection coefficients obtained from the full-wave simulations using the commercial software HFSS [25]. As can be seen, there are four resonances over the 0.8–2.2 GHz frequency band. They appear at frequencies 1.05, 1.55, 1.90, and 2.10 GHz, respectively. These resonant frequencies are in good agreement with those predicted by the cavity model.

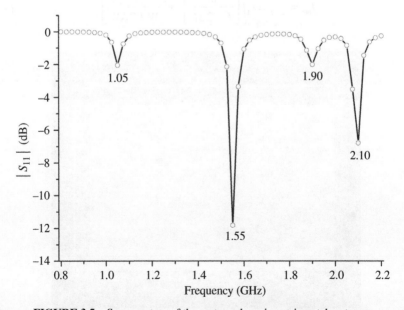

FIGURE 3.5   *S*-parameters of the rectangular microstrip patch antenna.

In order to examine the characteristics of the radiating currents of each resonant mode, Figure 3.6 plots the HFSS-simulated radiating currents at the resonant frequencies found from Figure 3.5. At the first resonant frequency 1.05 GHz, the rectangular patch antenna resonates along the length of the patch, which is the largest

(a)

(b)

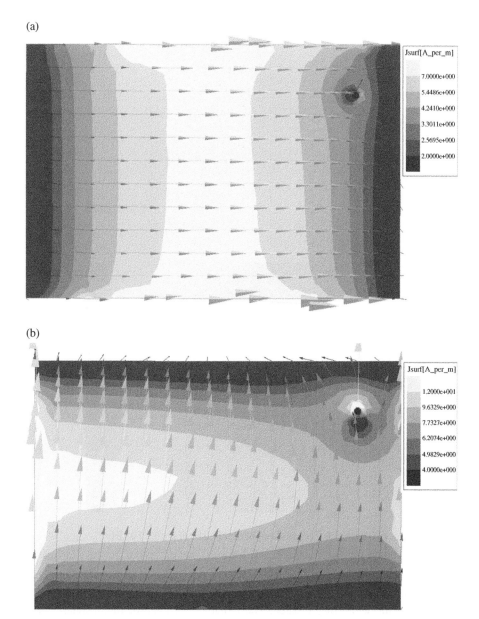

FIGURE 3.6   HFSS-simulated radiating currents at the resonant frequencies. (a) 1.05 GHz, (b) 1.55 GHz.

(c)

(d)

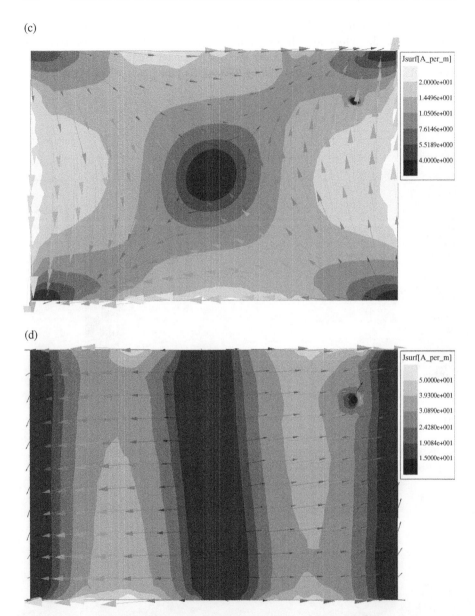

FIGURE 3.6   (*Continued*) (c) 1.90 GHz, and (d) 2.10 GHz.

dimension size of the patch. Therefore, it must be the lowest mode of the rectangular patch. At the second resonant frequency 1.55 GHz, the rectangular patch antenna resonates along the width of the patch. The radiating currents for this resonance are orthogonal with those at the first resonance frequency. In Figure 3.6c and d, radiating currents show that they have much shorter resonant current paths than the first two modes.

Therefore, they are higher order modes and resonate at much higher frequencies. In particular, Figure 3.6(d) shows that the fourth mode resonant along the half length of the patch, thus the resonant frequency is doubled from that of the first mode 1.05–2.10 GHz. This interesting observation is easy to understand from the point of view of the well-known antenna's half-wavelength resonant length.

Figure 3.7 shows the HFSS-simulated radiation patterns at the resonant frequencies. As can be observed from Figure 3.7a and b, both the first two resonances have broadside radiation patterns. They have larger beam widths in their E-planes than in the H-plane. The radiation patterns of these two resonances have orthogonal polarizations. The E-plane and H-plane at these two frequencies interchange with each other. This orthogonality is caused by the orthogonal radiating currents as shown in Figure 3.6a and b. In the radiation patterns of the resonances at 1.90 and 2.10 GHz, the maximum radiations appear at an oblique angle. As we shall see from the CM analysis, the natural modes of the rectangular patch govern the properties of the radiating currents and far-field radiation patterns. Consequently, we can obtain many different radiation patterns from the same radiating aperture by carefully designing the feeding structures for each resonant mode.

On the other hand, in the case that the feed point for the coaxial probe was not properly chosen, for example, placed at the center of the patch, we are impossible to obtain the radiating currents and radiation patterns as shown in Figures 3.6 and 3.7. In addition, it is also impossible to observe the resonant frequencies if the coaxial probe was not properly

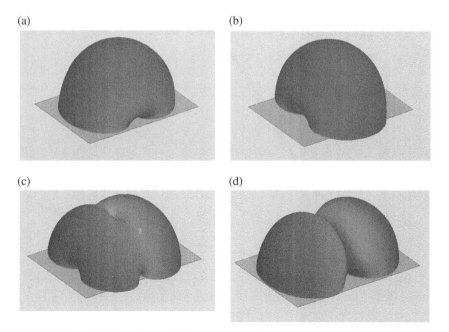

FIGURE 3.7   HFSS-simulated radiation patterns at the resonant frequencies. (a) 1.05 GHz, (b) 1.55 GHz, (c) 1.90 GHz, and (d) 2.10 GHz.

(a)

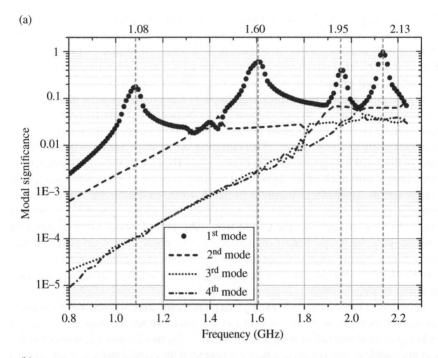

(b)

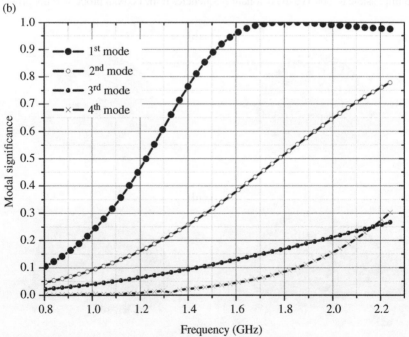

FIGURE 3.8   Modal significances obtained from (a) CM theory for antennas in multilayered medium and (b) CM theory for PEC objects.

placed. Fortunately, the CM theory for antennas in multilayered medium provides a source-free approach to investigate the resonant frequencies and radiation performance of microstrip patch antennas. The characteristic currents and fields are independent of the coaxial probe and its placement, and they are only dependent on the dielectric substrate, and the size and shape of the patch. Thus, the CM theory for antennas in multilayered medium brings many conveniences in the investigations of microstrip antennas.

Figure 3.8a shows the modal significances obtained from the CM theory for antennas in multilayered medium. According to the physical meaning of the modal significance, it can be observed from Figure 3.8a that there are four resonances over the 0.8–2.2 GHz frequency band. The predicted resonant frequencies for the TM_{10}, TM_{01}, TM_{11}, and TM_{20} modes locate at 1.08, 1.60, 1.95, and 2.13 GHz, respectively. These CM-predicted resonant frequencies agree well with those predicted by the cavity model and the full-wave simulations.

However, a review of previous literatures finds that CM analyses for microstrip patch antennas were usually carried out for isolated patches, that is, the dielectric material and ground plane, are neglected. As expected, such kind of simplification will affect the accuracy of the resonant frequencies. Figure 3.8b presents the modal significance of an isolated rectangular patch. It is computed using the CM theory for PEC objects. As can be seen, it is difficult to find the resonant frequency from Figure 3.8b. Considering the CM analysis results in Figure 3.8a, we can conclude that the ground plane and the dielectric substrates cannot be ignored in the seeking of the resonant frequencies for microstrip antennas.

Figure 3.9 displays the characteristic currents and characteristic fields of the dominant modes at the resonant frequencies. As compared to the full-wave simulation results in Figures 3.6 and 3.7, we can find that the CMs give more clear descriptions to the resonant behavior and radiation properties, although we have not considered any specific feeding structures. Nevertheless, both the figures illustrate the characteristics of radiating currents and radiation properties at each resonant frequency.

In summary, the resonant frequencies of the TM_{10}, TM_{01}, TM_{11}, and TM_{20} modes found by the CM theory for antennas in multilayered medium agree well with those predicted by the cavity model. Meanwhile, the characteristic currents and fields also agree well with the physically excited currents and fields obtained through the accurate full-wave simulations. It demonstrates that the CMs solved from the CM theory can be physically excited. It further illustrates that the CMs will be helpful in the microstrip patch antenna developments, in terms of both the feeding structure designs and patch shape and size optimizations. We will discuss the implementations of the CM theory in practical microstrip patch antenna designs in the following subsections.

3.4 PHYSICAL INVESTIGATIONS ON MICROSTRIP PATCH ANTENNAS

It has been demonstrated that the CM theory for antennas with multilayered medium provides clear physical insight into the resonant behavior of rectangular microstrip antennas. Although the CM theory is equivalent to the cavity model in the computation

(a)   (b)

(c)   (d)

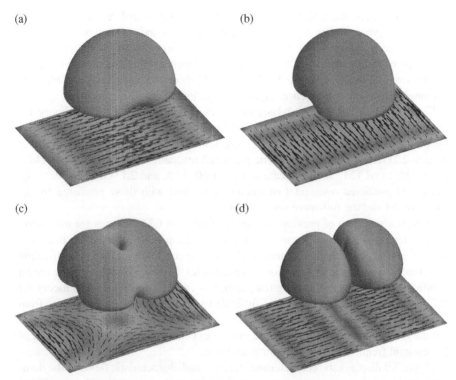

FIGURE 3.9   Characteristic currents and far fields of the dominant modes at the resonant frequencies. (a) TM_{10}, 1.08 GHz; (b) TM_{01}, 1.60 GHz; (c) TM_{11}, 1.95 GHz; and (d) TM_{20}, 2.13 GHz.

of the resonant frequencies, it is a more general theory and can be used to analyze arbitrarily shaped patches in multilayered medium. In this section, we will implement the developed CM theory in a variety of microstrip patch antennas. The CM analysis will reveal many interesting electromagnetic phenomenon and radiation mechanisms of these commonly used microstrip patch antennas.

3.4.1   Equilateral Triangular Patch Antenna

Equilateral triangular patch is a promising candidate in the design of single-feed CP antennas or compact antennas. This section examines the resonant behavior of an equilateral triangular patch antenna as shown in Figure 3.10 using the newly developed CM theory. The edge length of the equilateral triangular patch is 10 cm. It is printed on a dielectric substrate with a relative permittivity of $\varepsilon_r = 2.32$ and a thickness of $h = 1.6$ mm. This equilateral triangular patch antenna has also been investigated using the cavity model [9] and the said plane-wave incident method in Ref. [11]. Measurement results were also presented in Ref. [12] to validate the results computed from the plane-wave incident method.

Figure 3.11 shows the modal significances over the range 1.1–3.5 GHz. It can be clearly observed that there are seven resonances within the frequency band. Three

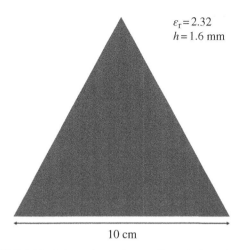

FIGURE 3.10   Dimensions of an equilateral triangular patch.

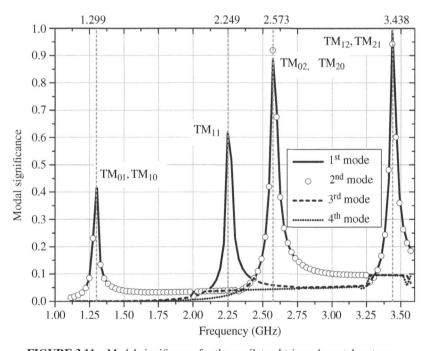

FIGURE 3.11   Modal significance for the equilateral triangular patch antenna.

pairs of them resonate at the same frequency, which exactly reflects the effect of the symmetry property in the equilateral triangular patch. Table 3.1 compares the resonant frequencies of the present method with those obtained from the cavity model [9], the plane-wave incident method [11], and the measurements [12]. As can be seen, the resonant frequencies from the three approaches agree well with each other.

TABLE 3.1   Resonant Frequencies Obtained From Different Methods

	Resonant frequency, GHz			
	Present method	Cavity model [9]	Plane-wave incident [11]	Measured results [12]
TM_{01}	1.299	1.299	1.249	1.280
TM_{10}	1.299	1.299	1.276	1.280
TM_{11}	2.249	2.252	2.172	2.242
TM_{02}	2.573	2.599	2.525	2.550
TM_{20}	2.573	2.599	2.510	2.550
TM_{12}	3.438	3.439	3.265	3.400
TM_{21}	3.438	3.439	3.356	3.400

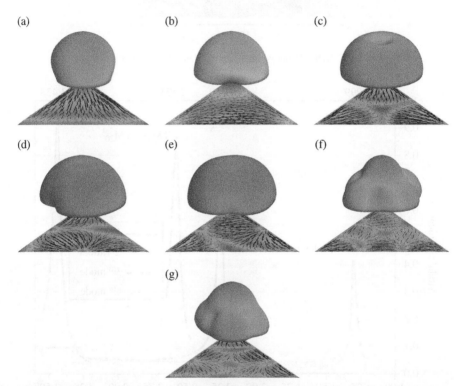

FIGURE 3.12   Characteristic currents and far-fields of the dominant modes at the resonant frequencies. (a) TM_{01}, 1.299 GHz; (b) TM_{10}, 1.299 GHz; (c) TM_{11}, 2.249 GHz; (d) TM_{02}, 2.573 GHz; (e) TM_{20}, 2.573 GHz; (f) TM_{21}, 3.438 GHz; and (g) TM_{12}, 3.438 GHz.

Figure 3.12 gives the characteristic currents and fields of the dominant CMs at each resonant frequency listed in Table 3.1. The mode pairs that resonate at the same frequency have different characteristic currents. Therefore, the characteristic currents should be considered together with the resonant frequencies to classify the different radiating modes. It is significant to note that the modal currents computed from the plane-wave incident method [11] are dependent on the incident field. Taking TM_{11} mode as an example, two different modal currents for TM_{11} mode are presented in Ref. [11]; they are actually the induced currents due to incident fields from different angles. Both of those TM_{11} mode currents are different from the characteristic currents in Figure 3.12c. Figure 3.12c illustrates the true resonant currents for the TM_{11} mode. By placing three probes at the current maxima near the three vertexes respectively, the current distribution shown in Figure 3.12c can be successfully excited at 2.249 GHz. However, we cannot excite the current distributions in Ref. [11] by placing any practical feeding structure. This is another explicit way to demonstrate the correctness of the characteristic currents computed from the CM theory.

It is well known that equilateral triangular patch antenna can provide CP waves by a single feed. A can be observed from Figure 3.12a and b, the characteristic currents for TM_{01} and TM_{10} modes can be identified as the vertical and horizontal polarized currents, respectively. The TM_{01} and TM_{10} mode are linearly polarized modes and they are orthogonal with each other. By properly exciting these two modes at the same time, we can achieve CP using the equilateral triangular patch. The higher order modes only resonant at some small portions of the patch, and their resonance frequencies are much higher than the fundamental TM_{01} and TM_{10} modes. Following the characteristic currents, we can excite these high-order modes at their resonant frequencies by carefully designing the feeding structures. They provide the design freedoms for multiband operations. As the characteristic fields of each mode are different from each other, we can also exploit the high-order modes for reconfigurable radiation pattern antennas designs.

In summary, the CM theory clearly indicates the underlying physics of equilateral triangular patch antenna. They are helpful in feeding structure designs for any possible radiation performance.

3.4.2 Concentric Circular Microstrip Antenna

Concentric circular microstrip antennas have received considerable attentions. They are often employed for beam scanning through weighting different radiating modes. Due to its multi-mode property, they are also popular in the reconfigurable radiation pattern antenna designs. This section discusses a multimode concentric circular microstrip patch antenna that is printed on a single dielectric substrate. Figure 3.13 shows the configuration of a concentric circular patch antenna. The concentric circular patch is printed on a dielectric substrate with dielectric constant $\varepsilon_r = 2.2$ and thickness $h = 0.8$ mm. The concentric circular patch consists of one center disk and two concentric rings. The radius of the center disk is $R1 = 11.3$ mm, and the outer radius of the two rings are $R2 = 16.8$ mm and $R3 = 24.8$ mm, respectively. The gap between the center disk and rings are $g = 0.5$ mm. The excitation of a variety of modes on the center disk and concentric rings through multiple feedings allows the reconfigurable radiation pattern antenna designs [26]. Beam scanning was also realized by combining different modes with suitable amplitude and phase excitations. However, full-wave simulations cannot give the modal currents and modal fields of individual radiating modes. The knowledge on how to excite and combine these modes has not been presented in an explicit way.

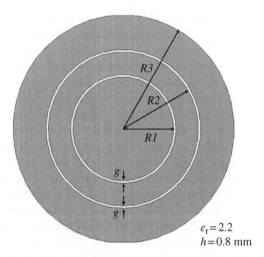

FIGURE 3.13   Configurations of the concentric circular patch antenna.

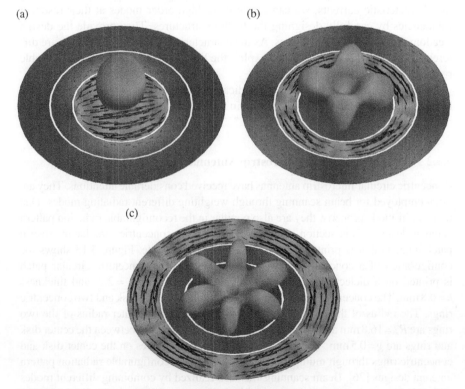

FIGURE 3.14   Characteristic currents and characteristic fields of the concentric circular patch at 5.0 GHz. (a) $\mathbf{J}_1$ & $\mathbf{E}_1$, (b) $\mathbf{J}_2$ & $\mathbf{E}_2$, and (c) $\mathbf{J}_3$ & E3.

Consequently, CM analysis is performed for this concentric circular patch antenna. Figure 3.14 shows the characteristic currents and characteristic fields of the first three modes at 5 GHz. As can be seen, the first mode mainly resonates on the center disk. The characteristic current of this mode is linear polarized and the corresponding characteristic field features broadside radiations. Both the characteristic current and characteristic fields show that this is essentially the TM_{11} mode of a circular patch antenna.

In the second mode, it is interesting to find that the radiating currents mainly appear on the first ring. There are four current peaks and four nulls along this ring. It illustrates that we can excite this mode by placing coaxial probe feedings at the locations with current peaks. Figure 3.14c illustrates that the third mode resonates on the second ring. Due to the increased number of current peaks and nulls on the second ring, there are six lobes in the characteristic fields. Similarly, the characteristic currents also indicate that we can excite this mode by placing coaxial probe feedings at the six current maximums.

As can be seen, the first three modes resonant over different parts of the concentric circular patch. It indicates that these three modes can be excited independently at the same frequency. This property allows us to excite these three modes simultaneously and may be possible to combine these modes with arbitrary complex excitation for a diversity of radiation patterns. Moreover, the characteristic currents also provide valuable information on the placement of feedings for the excitation of each mode. The designs reported in Ref. [26] demonstrate this concept through using multiple feedings on the center disk and the two rings.

3.4.3 Corner-Truncated Circularly Polarized Antenna

In situations when it is impossible to ascertain the polarization of an incoming wave, CP antennas are attractive candidates. This section discusses the CMs of a compact square CP antenna with truncated corners and insert slits. Figure 3.15 shows the configuration of the corner-truncated CP antenna as reported in Ref. [27]. The radiating patch is printed on a 1.6 mm thickness dielectric substrate, with a dielectric constant of $\varepsilon_r = 4.4$. Evidently, analytical methods such as the cavity models are not suitable to analyze such structures due to their complicated geometries.

By tuning the geometry parameters l and ΔL, the center frequency of the axial ratio (AR) band can be adjusted. Table 3.2 lists the geometry parameters for three corner-truncated CP antennas. Figure 3.16 shows the modal significances and characteristic angles for the three antennas with different slit lengths and truncated corners. As can be seen, at the resonant frequencies of each antenna, the first two modes have comparable modal significance and they are much larger than those of the third order and the other higher order modes. Therefore, these two modes work together at the center frequency for the CP operations. Meanwhile, in the three antennas, the phase differences of the characteristic angles for the first two modes are around 90°. Moreover, for each of the antenna, the 90° phase differences and the largest modal significances locate at the same frequency. Recalling the physical meanings of the modal significance and the characteristic angles, we can clearly see that the two dominant modes have the comparable current density and they generate the far fields in free

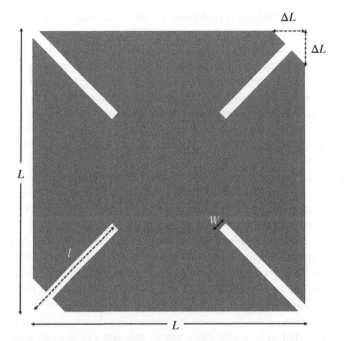

FIGURE 3.15   Configurations of the corner-truncated CP antenna.

TABLE 3.2   Geometry Parameters for the Three Corner-truncated CP Antennas

	L (mm)	w (mm)	l (mm)	ΔL (mm)
Antenna 1	28	1	16	6.3
Antenna 2	28	1	14	4.1
Antenna 3	28	1	12	3.3

space with a 90° phase differences. These are the radiation mechanisms found by the modal significance and characteristic angle for the interpretation of the CP radiations.

The characteristic currents of the first two dominant modes for antenna 1 are shown in Figure 3.17. It can be observed that the characteristic currents for the first two dominant modes are linearly polarized and are orthogonal with each other. This property is the necessary condition for CP antenna designs using single feed. Characteristic currents for antenna 2 and antenna 3 also hold the same properties. Figure 3.18 shows the characteristic fields of the first two dominant modes. It can be clearly observed that the far fields are orthogonal with each other in the infinite radiation sphere. These observations reveal that it is possible to achieve CP operations by exciting the two orthogonal modes simultaneously at the predicted frequencies as shown in Figure 3.16. This example shows well that the CM theory not only computes the resonant frequencies of each radiating modes accurately but it also gives clear physical interpretations to antenna's radiation mechanisms.

(a)

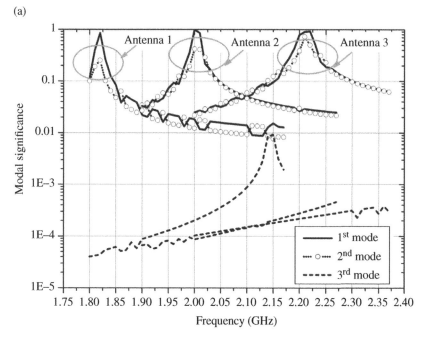

(b)

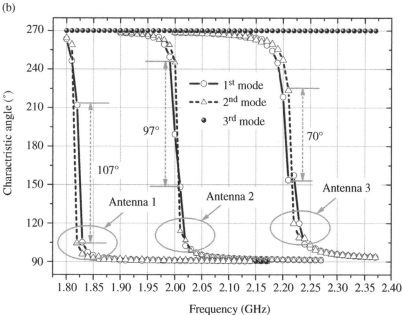

FIGURE 3.16 Modal significances (a) and characteristic angles (b) for the three corner-truncated CP antennas.

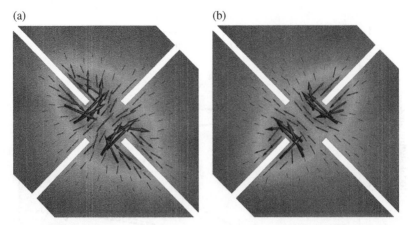

FIGURE 3.17   Characteristic currents for antenna 1: (a) mode 1 and (b) mode 2.

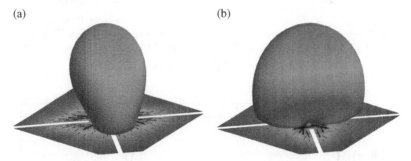

FIGURE 3.18   Characteristic fields for antenna 1: (a) mode 1 and (b) mode 2.

3.4.4   Dual Band Stacked Microstrip Patch Antenna

This section goes on to investigate the radiation mechanism of a dual band stacked microstrip antennas using the CM theory. Figure 3.19 shows configuration of the dual band circularly polarized stacked microstrip antenna reported in Ref. [28]. Two stacked corner-truncated square patches are printed on the top of two stacked dielectric substrates. The dielectric constant of the dielectric material is $\varepsilon_r = 4.4$, and each of them has a thickness of 1.6 mm. The two dielectric substrates are separated by a 0.45 mm thick air gap. The coaxial probe feed directly connect to the upper radiating patch through a via hole on the lower patch. The lower patch is thus excited through the electromagnetic couplings from the probe and the upper patch. The thickness of the thin air layer can be adjusted to vary the frequency ratio of the two operation bands. In Ref. [29], full-wave simulations and experimental validations were presented for this dual band stacked microstrip antenna. In this section, we first conduct the full-wave simulations to the stacked patches with consideration of the feeding probe. CM analysis is then carried out to the stacked patches in the multilayered medium environment. The obtained CM quantities, characteristic currents, and characteristic fields are presented to interpret the different radiation mechanism of the stacked patches in the two operation bands.

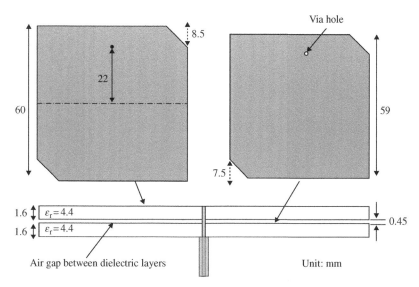

FIGURE 3.19   Configuration of the dual band stacked microstrip antenna.

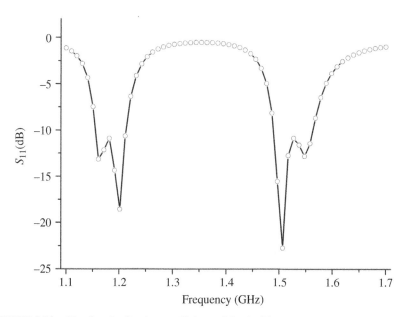

FIGURE 3.20   Simulated reflection coefficient of the dual band stacked microstrip antenna.

Figures 3.20 and 3.21 show the dual band operations in terms of the impedance matching and the AR performance, respectively. As can be seen, with the coaxial probe feed as shown in Figure 3.19, the stacked microstrip antenna has good input impedance matching and AR performance in the 1.2 and 1.5 GHz bands.

The excited current distributions on the stacked patches in the two bands are shown in Figure 3.22. Figure 3.22a shows that the currents on the lower patch dominate the

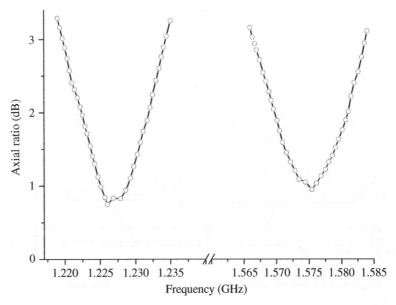

FIGURE 3.21 Measured axial ratio of the dual band stacked microstrip antenna in Ref. [29].

(a)

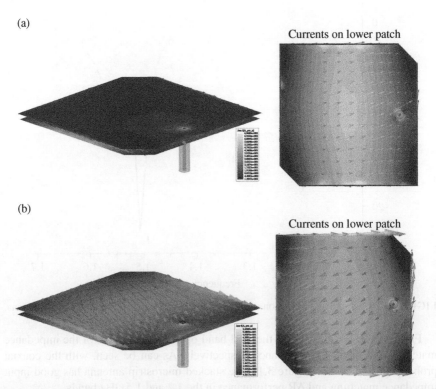

(b)

FIGURE 3.22 The current distributions on the stacked patches in two different bands: (a) 1.225 GHz band and (b) 1.575 GHz band.

radiations in the 1.2 GHz band. The currents on the upper patch are much weaker and contribute little to the radiations. As for the higher band, the current distributions in Figure 3.22b show that the currents on the lower and upper patches have comparable magnitudes. Therefore, both the patches contribute comparable radiations in the 1.5 GHz band.

Although the current distributions in Figure 3.22 illustrate some radiation mechanisms of the dual band stacked CP antenna, the premise is that the coaxial probe has to be placed correctly, such that the natural modes of the stacked patches are excited. With this premise, we can compute the current distributions through full-wave simulations and take some insights to operation mechanisms. However, we cannot get such physical insights before the accomplishment of the probe feed design and placement.

As has been discussed in the preceding sections, the CM theory is a source-free approach. It is able to provide the resonant behavior and radiation performance of an electromagnetic structure without considering any specific feeding structures and sources. The CM theory for antennas in multilayered medium is thus applied in the analysis of the stacked microstrip patch antenna. The via hole in the lower patch and the coaxial probe feed is not considered in the CM analysis.

Figure 3.23 shows the modal significance over the 1.1–1.7 GHz band. There are four dominant modes observed within the entire band. The first two modes have comparable modal significances in the 1.2 GHz band. Following the properties of the CMs, the first two modes are orthogonal with each other. Therefore, the two modes can be excited simultaneously with proper weightings for CP radiation. Similar phenomenon is observed in the 1.5 GHz band. The third and fourth modes have comparable modal significances in this high band. The orthogonality between these two modes also indicates that CP is possible by proper modal excitations.

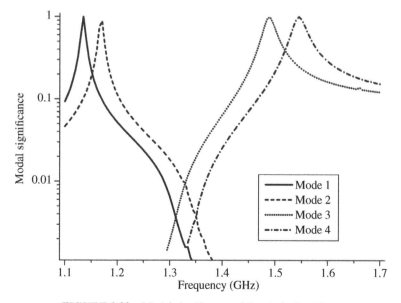

FIGURE 3.23 Modal significance of the stacked patches.

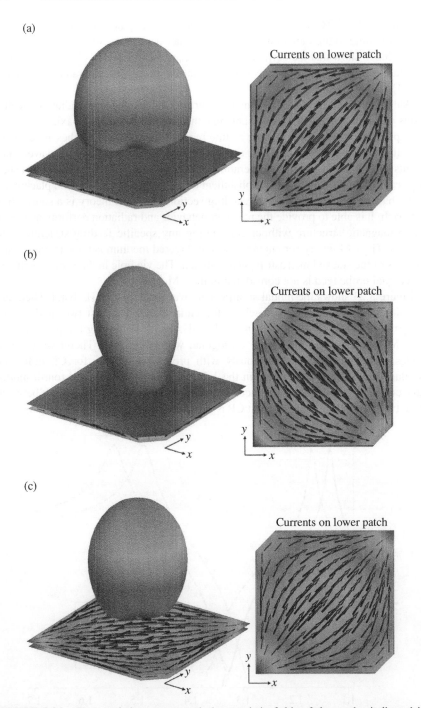

FIGURE 3.24 Characteristic currents and characteristic fields of the modes indicated in Figure 3.23. (a) The first mode at 1.15 GHz; (b) the second mode at 1.15 GHz; (c) the third mode at 1.5 GHz.

(d)

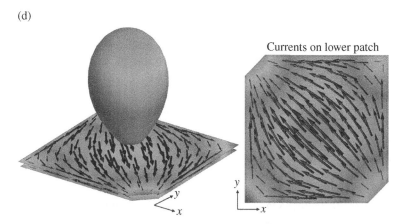

Currents on lower patch

FIGURE 3.24   (*Continued*) (d) the fourth mode at 1.5 GHz.

To further investigate the resonant behavior of the stacked microstrip patch antenna, Figure 3.24 displays the characteristic currents and characteristic fields for the four dominant modes indicated in Figure 3.23. As shown in Figure 3.24a and b, the first two modes in the lower band have orthogonal radiating currents on the lower patch, whereas the currents on the upper patch are much weaker. As compared to the lower patch, the upper patch contributes little to the radiation performances. Therefore, the lower patch dominates the radiation in the lower band. It should be noted that the same observation is obtained in the practically excited currents as shown in Figure 3.22a. Moreover, the characteristic fields of these two modes are also orthogonal with each other. These observations further illustrate that the first two modes can be exploited for CP radiations.

The characteristic currents and characteristic fields of the third and fourth modes in the 1.5 GHz bands are shown in Figure 3.24c and d, respectively. The characteristic currents on the lower and upper patches have comparable current magnitudes, which indicate that both the patches are in resonance state in the higher band. This observation is also the same as those obtained from the practically excited currents in Figure 3.22b. Moreover, the characteristic currents and characteristic fields of the two modes are also orthogonal with each other. It again indicates that circular polarization can be realized by properly exciting and combining these two modes. In summary, the CM theory provides an efficient approach for the analysis and understanding of dual band stacked microstrip patch antennas.

3.5   APPLICATIONS TO CIRCULARLY POLARIZED MICROSTRIP ANTENNA DESIGN

Circularly polarized microstrip antennas found wide applications in global positioning systems (GPS), mobile satellite, radio-frequency identification (RFID) design, and so on [28, 30, 31]. As discussed in many antenna textbooks [32, 33], the CP microstrip antennas can be viewed as the superposition of two linearly polarized modes with equal amplitude and quadrature phase excitations. The CM analysis in the preceding subsections also clearly demonstrates this operation principle of CP radiation. The various CP microstrip antennas differ primarily in how these linearly polarized modes are properly excited. Investigations have also shown that the most important issue associated with CP antenna designs is how to optimize the patch shapes and recognize the two linear modes to be excited for CP radiation.

Unfortunately, the current state of art for CP antenna design is always accomplished by tuning the patch shapes and feed positions through parameter sweeps or using automated optimization techniques [34–36]. However, these methods cannot ensure a final satisfactory solution in each design attempt. Even with the support of high-performance computers, the success of the final design still largely depends on the intuition and previous experience. Such experience is actually rather difficult to transfer from one to another. On the other hand, although various novel patch shapes are proposed to achieve CP, the major difficulty is that they are lacking physical insight. Real knowledge of the operating principles is usually difficult to understand, especially for those with complicated shapes.

In this section, CM analysis is employed as an efficient way to gain physical insights into the fundamental electromagnetic properties of two typical single feed CP patch antennas, that is, U-slot [37] and E-shaped patch antenna [38]. Based on CM theory, we present a framework of CM analysis to figure out the orthogonal modes and operating frequency band for potential good AR performance [39]. We also discover the physical factors that lead to high cross polarization level. Approach based on CM analysis results is also given to determine the optimal feed position [39]. Moreover, the representative CP antenna designs show that an offset probe feed will provide better AR performance for CP U-slot patch antennas. It also shows that cutting off the redundant section on the CP E-shaped patch will yield low cross polarization and more compact antenna size.

3.5.1 U-Slot Microstrip Antenna

This section discusses the application of the CM theory to find out the optimal feeding point for good AR performance in a U-slot microstrip antenna. Figure 3.25 shows the geometry and prototype of the proposed CP U-slot antenna. As compared to the U-slot antenna in Ref. [37], the probe feed is moved from the center to the inner edge of the U-slot's long arm. The radiating patch was fabricated using 0.5 mm thick copper sheet, and the ground plane was fabricated on a 1.6 mm thick FR4 substrate.

CM analysis was first carried out for the U-slot patch over a frequency band of 2.2–2.7 GHz. CMs are independent of excitation, and they are only dependent on the shape of the radiating patch. Therefore, the feeding structure is not considered in the CM analysis. For this electrically small problem, the first two CMs are sufficient to describe the nature of resonant behavior of the antenna. All the other modes can be viewed as higher order modes that are difficult to excite in practice. Figure 3.26 shows the characteristic angle and the modal significance of the first two modes. It is found that the two modes have a 90° phase difference at 2.3 GHz. Meanwhile, they have the exactly same modal significance at 2.3 GHz.

(a)   (b)

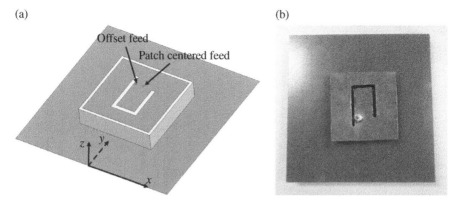

FIGURE 3.25 The circularly polarized U-slot antenna with offset feed: (a) the geometry and (b) prototype. From Ref. [39]. © 2012 by IEEE. Reproduced by permission of IEEE.

Figure 3.27 shows the characteristic currents of the U-slot patch. As shown in Figure 3.27a, the mode $\mathbf{J}_1$ can be characterized as the horizontal mode, with intense currents concentrating at the top and bottom section of the patch. It is also evident in Figure 3.27b that the mode $\mathbf{J}_2$ can be identified as the vertical mode, with its intense currents concentrating along the two arms of the U-slot. Ideally, good CP performance will appear at 2.3 GHz when these two modes could be excited and combined properly. The rest work is just to find the optimal probe feed position using the information in the characteristic currents.

As discussed above, it will be desirable if a single probe feed can excite the two orthogonal modes properly. In order to investigate the optimal area probe positions, the vertical mode is subtracted from the horizontal mode. Valuable information is obtained for optimal probe placement. Figure 3.27c shows the current distribution after subtraction, we refer to it as the "H-V" current $\mathbf{J}_1 - \mathbf{J}_2$ for convenience. It is interesting to note that the minimum current area locates at the inner edge of the U-slot's long arm, which indicates that the two modes present almost the same current amplitude. Therefore, with the expectation of good CP performance, we move the originally patch centered probe to the inner edge of the U-slot's long arm, as we have shown in Figure 3.25.

Figure 3.28 compares the measured and HFSS-simulated AR at the broadside direction ($\vartheta = 0°$, $\varphi = 0°$). Clearly, both the measured and HFSS-simulated results show that there is a significant improvement in the AR when the probe is moved to the inner edge of the U-slot's long arm. Furthermore, as expected from the CM analysis, the best AR performance is found nearly at 2.3 GHz. Figure 3.29a and b compare the measured left-hand circular polarization (LHCP) and right-hand circular polarization (RHCP) patterns with those obtained from HFSS simulations in the $\varphi = 0°$ and $\varphi = 90°$ planes. As can be seen, the measured patterns agree fairly well with the HFSS simulation results.

This example implies that there exists an area where the probe can be placed to achieve good AR performance. The optimal feed area for CP U-slot patch antenna is

(a)

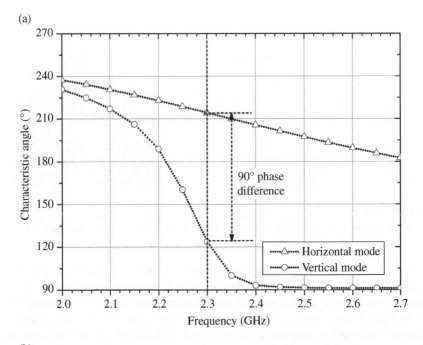

(b)

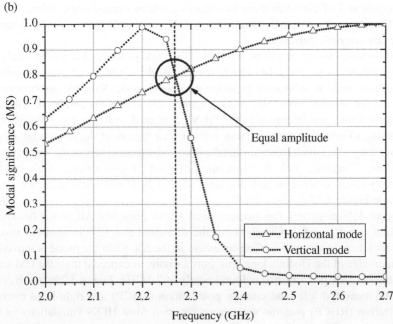

FIGURE 3.26 Horizontal and vertical modes of the U-slot antenna: (a) characteristic angle and (b) modal significance.

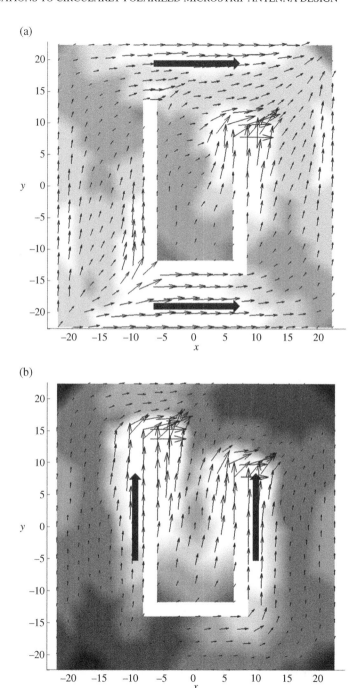

FIGURE 3.27   Characteristic currents of the U-slot antenna at 2.3 GHz. (a) Normalized horizontal mode $\mathbf{J}_1$, (b) normalized vertical mode $\mathbf{J}_2$.

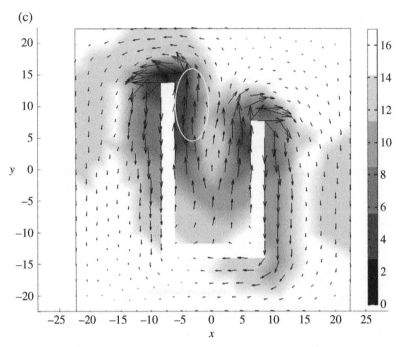

FIGURE 3.27   (*Continued*) (c) "H-V" current $\mathbf{J}_1 - \mathbf{J}_2$. The color scale is in dB and is applicable to all the three figures.

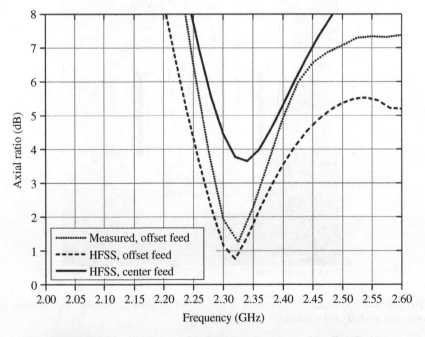

FIGURE 3.28   The axial ratio of the U-slot antenna with offset feed.

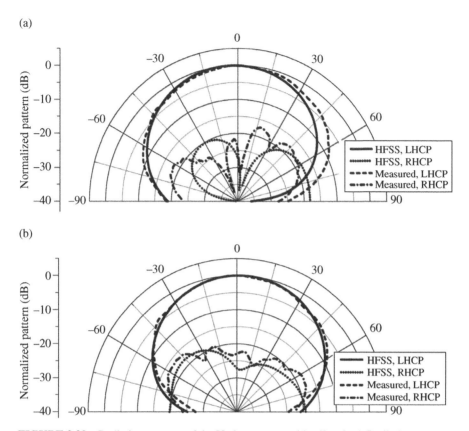

FIGURE 3.29 Radiation patterns of the U-slot antenna with offset feed. Radiation patterns at (a) 2.325 GHz in the $\phi = 0°$ plane and (b) 2.325 GHz in the $\phi = 90°$ plane.

at the inner edge of the U-slot's long arm, rather than the patch centered area. Both the experimental and HFSS-simulated results validate the improvement in AR performance. As such, the offset feed configuration can be taken as a rule of thumb for further CP U-slot patch antenna development.

3.5.2 E-Shaped Microstrip Antenna with Low Cross Polarization

In this section, we will discuss the application of the CM theory in CP E-shaped patch antenna optimization. It will illustrate how to suppress the cross polarization based on the CM analysis of the radiating patch. It is well known that E-shaped patches have gained a lot of popularity in widening impedance bandwidth [40]. On the other hand, by tuning the depths of the two slots, an unsymmetrical E-shaped patch is proposed by Khidre et al. to generate circularly polarized radiation [38]. The geometry and prototype of the unsymmetrical E-shaped patch antenna are shown in Figure 3.30. A foam with dielectric constant of $\varepsilon_r = 1.1$ is used to support the E-shaped patch. All the dimensions are kept the same as those of the prototype in

(a)

(b)

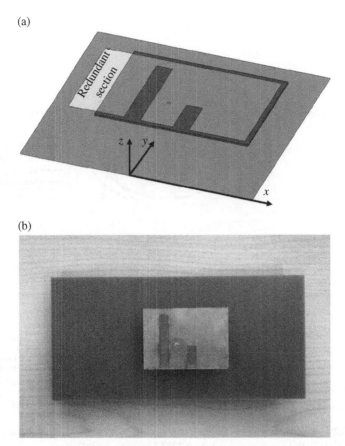

FIGURE 3.30 The circularly polarized E-shaped patch antenna with reduced size: (a) the geometry and (b) prototype. From Ref. [39]. © 2012 by IEEE. Reproduced by permission of IEEE.

Ref. [38]. As observed from the prototype, the redundant section is cut off from the one shown in Figure 3.30a, and we will later show that the size-reduced antenna will have enhanced polarization purity performance.

Figure 3.31 shows the characteristic angle and modal significance variations with frequency for the first three modes. As can be observed, mode $\mathbf{J}_1$ and $\mathbf{J}_2$ have the same modal current magnitude at 2.3 GHz, while mode $\mathbf{J}_2$ and $\mathbf{J}_3$ have the same current magnitude at 2.6 GHz. In addition, according to the physical interpretations of the CMs, it is readily seen that mode $\mathbf{J}_3$ and all the other higher order modes (they have lower modal significance than that of $\mathbf{J}_3$) contribute very little radiated energy at 2.3 GHz for any external excitation because their associated modal significance (<0.1) clearly indicates the difficulty to excite them.

Figure 3.32 shows the normalized current distribution for mode $\mathbf{J}_1$, $\mathbf{J}_2$, and $\mathbf{J}_3$ at the frequencies of interest. It is observed that mode $\mathbf{J}_1$ and $\mathbf{J}_2$ are characterized by

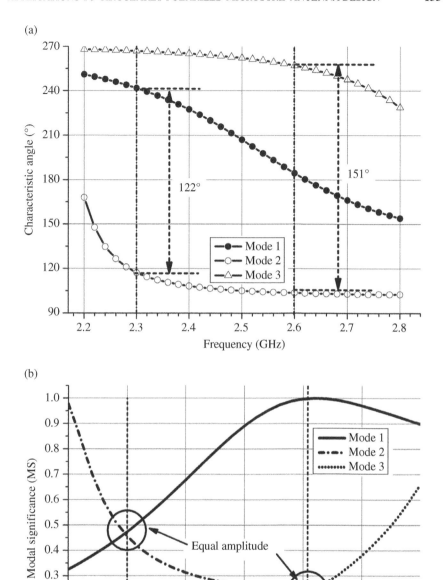

FIGURE 3.31 The first three characteristic modes of the E-shaped patch antenna: (a) characteristic angle and (b) modal significance.

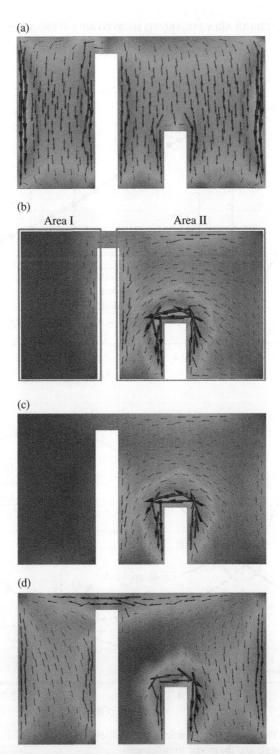

FIGURE 3.32 Characteristic current distribution of the E-shaped patch antenna. (a) Mode $\mathbf{J}_1$ at 2.3 GHz, (b) mode $\mathbf{J}_2$ at 2.3 GHz, (c) mode $\mathbf{J}_2$ at 2.6 GHz, and (d) mode $\mathbf{J}_3$ at 2.6 GHz. From Ref. [39]. © 2012 by IEEE. Reproduced by permission of IEEE.

vertical and horizontal currents, respectively, and $\mathbf{J}_3$ is also a vertical current. Together with the information provided in Figure 3.31, it is readily seen that $\mathbf{J}_1$ and $\mathbf{J}_2$ have the potential to be combined for CP near 2.3 GHz, and mode $\mathbf{J}_2$ and $\mathbf{J}_3$ can be properly combined to generate CP near 2.6 GHz. However, since mode $\mathbf{J}_1$ is the dominant mode at 2.6 GHz, it does not contribute to the CP, the cross-polarization level should be greater than that at 2.3 GHz. Therefore, even the feeding structure has not been considered yet; we can conclude from the CM analysis that good CP performance can be obtained within a narrow frequency band near 2.3 GHz, but it is difficult to achieve good CP at 2.6 GHz. This is one of the most important features that we can benefit from the CM theory and it is also what makes the CM analysis be different from other traditional full-wave analysis.

This property is indeed helpful for practical design, because one cannot get a satisfactory design if the patch shape and size has not been properly designed, regardless of what the feed configuration is. The discussion we have presented can directly lead to a "two-step" scheme for circularly polarized antenna design. The first step is to determine the patch shape and size such that it has two orthogonal modes with 90° characteristic angle difference and similar modal significance. The second step is to find an optimum feed position using the method we have introduced in the U-slot antenna design.

Moreover, observation of Figure 3.32b illustrates that the current distribution in Area I is much weaker than that in Area II. Thus horizontal mode dominantly supporting the CP is mainly contributed from the current in Area II. In addition, Figure 3.32a shows that the vertical mode current in Area I is comparable to that in Area II. It is easy to learn that Area I is the redundant section that does not contribute to CP and will inevitably rise up the cross-polarization level. Therefore, with the expectation of reducing the high cross-polarization level (approximately −10 dB in the range ±30° from broadside) as described in Ref. [38], we cut the redundant section shown in Figure 3.30a. The characteristic currents for the size-reduced E-shaped patch antenna are shown in Figure 3.33. As can be observed, this cutting does not affect the vertical mode, but the horizontal mode becomes more evident as compared with the one in Figure 3.32b. This fact demonstrates the feasibility of the redundant section cutting again.

The reflection coefficients of the circular polarized E-shaped patch antenna are shown in Figure 3.34. It is clearly seen that this cutting does not greatly affect the reflection coefficient performance. The reason is that the currents on the redundant section are very weaker. It does not contribute to the radiations in the desired band. On the other hand, Figure 3.35 shows that the AR between 2.7 and 2.8 GHz seems to degrade much. However, from Ref. [38] we can see that the CP gain in this frequency range is 5–7.5 dB lower than that at 2.3 GHz. It is also well known that if the AR performance is as good as those at low frequency, the gain at higher frequency band should be larger. Moreover, in the frequency band of 2.7–2.8 GHz, the 3 dB LHCP beamwidth is quite narrow, and the RHCP level is comparable to the LHCP level, although the AR at the broadside direction is fairly good. From this point of view, we can recognize that the redundant section cutting does not degrade the CP performance in the effective frequency range that has good impedance matching.

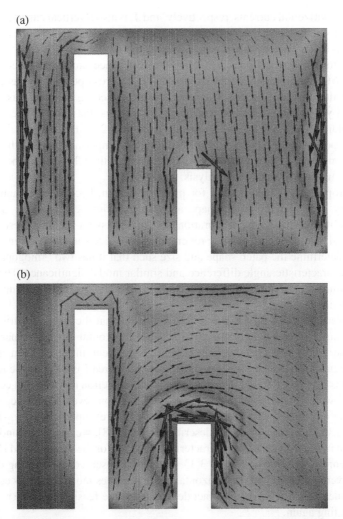

FIGURE 3.33 Characteristic current distribution of the size-reduced E-shaped patch antenna at 2.3 GHz. (a) Mode $\mathbf{J}_1$ and (b) mode $\mathbf{J}_2$. From Ref. [39]. © 2012 by IEEE. Reproduced by permission of IEEE.

Figure 3.36 shows the radiation patterns of the circular polarized E-shaped patch antenna. As can be seen, both the measured and simulated radiation patterns show that the cross-polarization level of the size-reduced E-shaped patch antenna keeps below −15 dB in the angular range ±90° from broadside, which is a significant improvement as compared with those reported in Ref. [40].

This design illustrates that the redundant section that does not contribute to CP can be easily identified with the aid of CMs. The final CP E-shaped patch antenna shows that the E-shaped patch is not necessary to be symmetrical with the probe. By cutting off the redundant section on the long slit side, CP antennas with lower cross

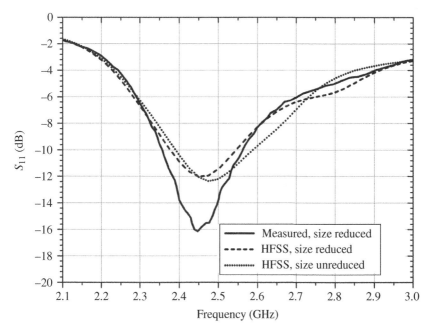

FIGURE 3.34   Reflection coefficients of the circular polarized E-shaped patch antenna.

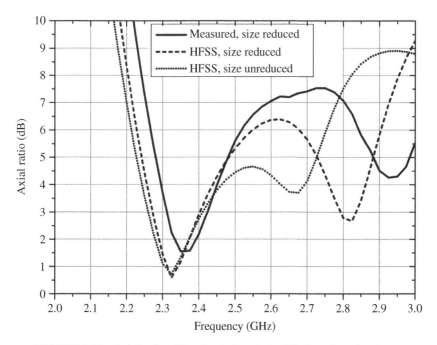

FIGURE 3.35   Axial ratio of the circularly polarized E-shaped patch antenna.

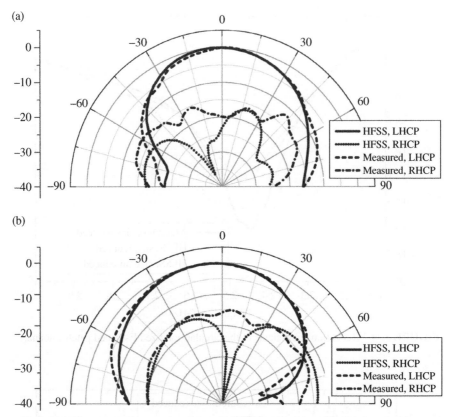

(a)

(b)

FIGURE 3.36 Radiation patterns of the circularly polarized E-shaped patch antenna at 2.45 GHz: (a) the $\phi=0°$ plane and (b) the $\phi=90°$ plane.

polarization can be obtained. As compared with traditional "cut and try" methods, the CM theory provides an explicit way with clear and deep physical insights for the optimization of CP antennas.

3.5.3 Summary

This section demonstrates the application of the CM theory for circularly polarized microstrip antenna designs. Optimal designs of the two typical CP patch antennas demonstrate that CMs provide invaluable information for performance improvement in AR and polarization purity. Experimental and simulated results are presented to validate the effectiveness of the CP antenna optimization method based on the CM analysis. On the other hand, the two CP microstrip antenna designs illustrate that (1) an offset probe feed in CP U-slot patch antennas will provide excellent AR performance and (2) a redundant section exists on the classic CP E-shaped patch antenna. Lower cross polarization can be obtained by cutting off this redundant section.

3.6 CONCLUSIONS

This chapter begins with a review for the state-of-the-art design approaches for microstrip patch antennas. The challenges in the traditional "cut and try" approaches are also discussed. It motivates the development of the CM theory for structures buried in multilayered medium. The theoretical formulations for the CM theory of structures in multilayered medium are then followed. With this newly developed CM theory, this chapter investigates many commonly used patch antennas and discusses their radiation mechanisms through the point of view of CMs. These investigated antennas include the rectangular patch antenna, triangular patch antenna, concentric circular patch antenna, corner cut circularly polarized antenna, and dual band stacked microstrip antennas. Many interesting observations and valuable information is obtained. They are helpful for enhancing further the radiation performance and also feeding structure designs. Furthermore, accurate full-wave simulation results and/or experimental results for these antennas are also presented to demonstrate that the CMs can be practically excited through properly designed feedings. Eventually, we illustrate the applications of the CM theory in two circularly polarized antenna designs: U-slot and E-shaped CP patch antenna designs. These two designs show that the CMs provide valuable information for the feeding placement, patch shape and size optimization, AR enhancement, and cross polarization suppression.

REFERENCES

[1] G. Deschamps and W. Sichak, "Microstrip Microwave Antennas," in Proceedings of the Third Symposium on the USAF Antenna Research and Development Program, Monticello, IL, Oct. 18–22, 1953.

[2] D. M. Pozar and D. H. Schaubert, *Microstrip Antennas: The Analysis and Design of Microstrip Antennas and Arrays*. Hoboken, NJ: John Wiley & Sons, Inc., 1995.

[3] K. L. Wong, *Design of Nonplanar Microstrip Antennas and Transmission Lines*. New York, NY: John Wiley & Sons, Inc., 1999.

[4] G. Garg, P. Bhartia, I. Bahl, and A. Ittipiboon, *Microstrip Antenna Design Handbook*. Norwood, MA: Artech House, 2002.

[5] K. L. Wong, *Compact and Broadband Microstrip Antennas*. New York, NY: John Wiley & Sons, Inc., 2002.

[6] G. Kumar and K. P. Ray, *Broadband Microstrip Antennas*. Norwood, MA: Artech House, 2003.

[7] K. L. Wong, *Planar Antennas for Wireless Communications*. Hoboken, NJ: John Wiley & Sons, Inc., 2003.

[8] Z. N. Chen and M. Y. W. Chia, *Broadband Planar Antennas*. Sussex, England: John Wiley & Sons, Ltd., 2006.

[9] K.-F. Lee, K.-M. Luk, and J. S. Dahele, "Characteristics of the equilateral triangular patch antenna," *IEEE Trans. Antennas Propag.*, vol. 36, no. 11, pp. 1510–1518, Nov. 1988.

[10] H. Pues and A. V. Capelle, "Accurate transmission-line model for the rectangular microstrip antenna," *IEE Proc.*, vol. 131, no. 6, Dec. 1984.

[11] M. D. Deshpande, D. G. Shively, and C. R. Cockrell, "Resonant frequencies of irregularly shaped microstrip antennas using method of moments," NASA Technical paper 3386, Hampton, VA, Oct. 1993.

[12] M. D. Deshpande, C. R. Cockrell, F. B. Beck, and C. J. Reddy, "Analysis of electromagnetic scattering from irregularly shaped, thin, metallic flat plates," NASA Technical paper 3361, Hampton, VA, Sep. 1993.

[13] G. Xu, "On the resonant frequencies of microstrip antennas," *IEEE Trans. Antennas Propag.*, vol. 37, no. 2, pp. 245–247, Feb. 1989.

[14] R. Harrington and J. Mautz, "Theory of characteristic modes for conducting bodies," *IEEE Trans. Antennas Propag.*, vol. 19, no. 5, pp. 622–628, Sep. 1971.

[15] J. Eichler, P. Hazdra, M. Capek, and M. Mazanek, "Modal resonant frequencies and radiation quality factors of microstrip antennas," *Int. J. Antennas Propag.*, vol. 2012, p. 9, 2012.

[16] C. V. Niekerk and J. T. Bernhard, "Characteristic mode analysis of a shorted microstrip patch antenna," in IEEE Antennas Propagation Society International Symposium, Chicago, IL, pp. 1–2, 2012.

[17] M. Cabedo-Fabres, E. Antonio-Daviu, M. Ferrando-Bataller, and A. Valero-Nogueira, "On the use of characteristic modes to describe patch antenna performance," in IEEE Antennas Propagation Society International Symposium, Columbus, OH, pp. 712–715, 2003.

[18] D. G. Fang, J. J. Yang, and G. Y. Delisle, "Discrete image theory for horizontal electric dipoles in a multilayered medium," *IEE Proc.*, vol. 135, no. 5, pp. 297–303, Oct. 1988.

[19] Y. L. Chow, J. J. Yang, D. G. Fang, and G. E. Howard, "A closed-form spatial Green's function for the thick microstrip substrate," *IEEE Trans. Microw. Theory Tech.*, vol. 39, no. 3, pp. 588–592, Mar. 1991.

[20] C. F. Wang, F. Ling, and J. M. Jin, "A fast full-wave analysis of scattering and radiation from large finite arrays of microstrip antennas," *IEEE Trans. Antennas Propag.*, vol. 46, pp. 1467–1474, Oct. 1998.

[21] F. Ling, C. F. Wang, and J. M. Jin, "An efficient algorithm for analyzing large-scale microstrip structures using adaptive integral method combined with discrete complex image method," *IEEE Trans. Microw. Theory Tech.*, vol. 48, pp. 832–839, May, 2000.

[22] F. Ling, J. Liu, and J. M. Jin, "Efficient electromagnetic modeling of three-dimensional multilayer microstrip antennas and circuits," *IEEE Trans. Microw. Theory Tech.*, vol. 50, no. 6, pp. 1628–1635, Jun. 2002.

[23] Y. T. Lo, D. Solomon, and W. F. Richards, "Theory and experiment on microstrip antennas," *IEEE Trans. Antennas Propag.*, vol. AP-27, no. 2, pp. 137–145, Mar. 1979.

[24] W. F. Richards, Y. T. Lo, and D. D. Harrison, "An improved theory for microstrip antennas and applications," *IEEE Trans. Antennas Propag.*, vol. AP-29, no. 1, pp. 38–46, Jan. 1981.

[25] ANSYS Corporation. Homepage of HFSS. Available at http://www.ansys.com/Products/Simulation+Technology/Electronics/Signal+Integrity/ANSYS+HFSS, Sep. 2014. Accessed 27 January, 2015.

[26] T. Q. Tran and S. K. Sharma, "Radiation characteristics of a multimode concentric circular microstrip patch antenna by controlling amplitude and phase of modes," *IEEE Trans. Antennas Propag.*, vol. 60, no. 3, pp. 1601–1605, Mar. 2012.

[27] W.-S. Chen, C.-K. Wu, and K.-L. Wong, "Novel compact circularly polarized square microstrip antenna," *IEEE Trans. Antennas Propag.*, vol. 49, no. 3, pp. 340–342, Mar. 2001.

[28] H. Iwasaki, "A circularly polarized small-size microstrip antenna with a cross slot," *IEEE Trans. Antennas Propag.*, vol. 44, no. 10, pp. 1399–1401, Oct. 1996.

[29] C.-M. Su and K.-L. Wong, "A dual-band GPS microstrip antenna," *Microw. Opt. Technol. Lett.*, vol. 33, no. 4, pp. 238–240, May 2002.

[30] K. R. Carver and J. W. Mink, "Microstrip antenna technology," *IEEE Trans. Antennas Propag.*, vol. 29, no. 1, pp. 2–24, Jan. 1981.

[31] K. L. Wong, C. C. Huang, and W. S. Chen, "Printed ring slot antenna for circular polarization," *IEEE Trans. Antennas Propag.*, vol. 50, no. 1, pp. 75–77, Jan. 2002.

[32] C. A. Balanis, *Antenna Theory: Analysis and Design* (3rd edition). Hoboken, NJ: John Wiley & Sons, Inc., pp. 75–76, 2005.

[33] C. A. Balanis, *Modern Antenna Handbook*. Hoboken, NJ: John Wiley & Sons, Inc., pp. 30–31, 2008.

[34] E. E. Altshuler, "Design of a vehicular antenna for GPS/Iridium using a genetic algorithm," *IEEE Trans. Antennas Propag.*, vol. 48, no. 6, pp. 968–972, Jun. 2000.

[35] R. Zentner, Z. Sipus, and J. Bartolic, "Optimization synthesis of broadband circularly polarized microstrip antennas by hybrid genetic algorithm," *Microw. Opt. Technol. Lett.*, vol. 31, no. 3, pp. 197–201, Nov. 2001.

[36] R. L. Haupt, "Antenna design with a mixed integer genetic algorithm," *IEEE Trans. Antennas Propag.*, vol. 55, no. 3, pp. 577–582, Mar. 2007.

[37] K. F. Tong and T. P. Wong, "Circularly polarized U-slot antenna," *IEEE Trans. Antennas Propag.*, vol. 55, no. 8, pp. 2382–2385, Aug. 2007.

[38] A. Khidre, K. F. Lee, F. Yang, and A. Eisherbeni, "Wideband circularly polarized E-shaped patch antenna for wireless applications," *IEEE Antennas Propag. Mag.*, vol. 52, no. 5, pp. 219–229, Oct. 2010.

[39] Y. Chen and C.-F. Wang, "Characteristic-mode-based improvement of circularly polarized U-slot and E-shaped patch antennas," *IEEE Antennas Wirel. Propag. Lett.*, vol. 11, pp. 1474–1477, 2012.

[40] F. Yang, X. X. Zhang, X. Ye, and Y. Rahmat-Samii, "Wide-band E-shaped patch antennas for wireless communications," *IEEE Trans. Antennas Propag.*, vol. 49, no. 7, pp. 1094–1100, Jul. 2001.

4

CHARACTERISTIC MODE THEORY FOR DIELECTRIC RESONATORS

4.1 BACKGROUNDS

Dielectric resonators were initially developed as high Q circuit elements, such as filters and oscillators [1]. In those circuit applications, dielectric resonators are normally fabricated with high permittivity material. The Q value can be as high as 10,000. Therefore, dielectric resonators were treated as energy storage elements in the early days. The resonant modes in dielectric resonators were first analyzed in 1962 by Okaya and Barash [2]. Although the dielectric resonators were occasionally found to radiate energy into free space in the circuit applications, the idea of intentionally exciting the appropriate radiating modes for antenna applications had not been widely recognized until the early 1980s [3–5]. In the pioneering work [3–5], the investigations into the resonant frequency, the Q value, the modal radiation patterns, and corresponding excitation schemes demonstrated that dielectric resonators could be designed into antennas. Dielectric resonators for antenna uses are later known as the dielectric resonator antennas (DRA). It has become an attractive alternative to traditional low-gain microstrip antennas. Owing to the elimination of the severe conductor loss and very low material loss in the millimeter wave frequency range (100–300 GHz), DRA has also been the promising candidate for millimeter wave antennas [6–9]. In the current context, the dielectric resonators refer to those designed as DRAs.

Characteristic Modes: Theory and Applications in Antenna Engineering, First Edition.
Yikai Chen and Chao-Fu Wang.
© 2015 John Wiley & Sons, Inc. Published 2015 by John Wiley & Sons, Inc.

4.1.1   A Brief Introduction to DRA

DRAs can be generally in the form of any three-dimensional shapes. As compared to the very popular microstrip patch antenna, DRA offers high degree of design flexibilities for a wide range of physical and electrical requirements in various communication systems. The primary characteristics of DRA can be summarized in the following,

- Compact size. The size of the DRA is proportional to wavelength in the dielectric material $\lambda_0 / \sqrt{\varepsilon_r}$, where λ_0 is the wavelength in free space and ε_r is the dielectric constant of the dielectric material. ε_r is usually larger than 7.0 and can be as high as 100 for more compact DRA size [10, 11]. Therefore, DRAs usually are in compact size.
- High-radiation efficiency can be achieved over a wide range of frequency band by choosing low-loss dielectric material, even in the millimeter wave frequency range.
- For the ease of integration into various communication systems, plenty of excitation methods are available that can be implemented for the feeding of the DRAs.
- More than one single mode can be excited simultaneously to obtain broadside radiation pattern and omnidirectional radiation patterns at different frequencies. It is also possible to excite many modes to achieve dual-band, multiband, or even broadband performances [12].

The most widely used DRAs include the cylindrical DRA, hemispherical DRA, and rectangular DRA. Because of the simple geometry of these commonly used DRAs, they are easy to fabricate in mass production. Figure 4.1 shows these simple DRA shapes with three typical excitation methods: the coaxial probe feeding, the aperture coupling feeding, and the microstrip line feeding.

As shown in Figure 4.1a, the probe feeding consists of the inner conductor of a coaxial cable that extends through the ground plane. From the point of view of equivalent source, the probe feeding can be considered as a vertical electric current source. To achieve strong coupling with a particular mode of a DRA, the probe feeding has to be placed in the region with strong electric fields. In this sense, the internal electric field within the DRA is of great importance in the placement of the probe. The amount of coupling between the probe and the DRA can be optimized by adjusting the probe height and the location with respect to the DRA. Because the region having strong electrical field varies with different modes, different probe location results in the excitation of different radiating modes in the DRA. Rigorous analyses for probe-fed hemispherical DRA and cylindrical DRA were carried out in Refs. [13–15]. The effects of the probe position and length on the input impedance were carefully investigated. However, there is no simple equation to determine the optimal probe length and location for a particular DRA with given dimensions and dielectric permittivity.

Figure 4.1b shows the configuration of an aperture coupling–fed rectangular DRA. In the aperture coupling feeding, the electromagnetic energy is coupled from the energy leaked out of the slot in the ground plane. The slot is indirectly fed with a

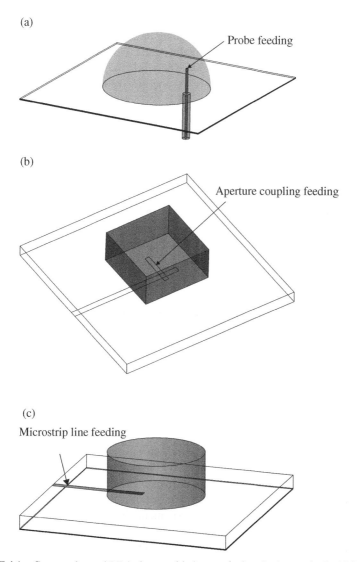

(a)

Probe feeding

(b)

Aperture coupling feeding

(c)

Microstrip line feeding

FIGURE 4.1 Commonly used DRA shapes with three typical excitation methods. (a) Probe-fed hemispherical DRA, (b) aperture coupling–fed rectangular DRA, and (c) microstrip line–fed cylindrical DRA.

microstrip line on the other side of the dielectric substrate. The configuration of the aperture coupling offers the advantage of having the feeding lines below the ground plane. The ground plane isolates the DRA radiator from the possible unwanted spurious radiations from the feeding line. Owing to these attractive features, the aperture coupling feeding has been widely adopted in many DRA designs, such as aperture coupling–fed rectangular and triangular DRAs [16], aperture coupling–fed cylindrical DRA [17], and aperture coupling–fed split cylindrical DRA [18].

The aperture coupling feeding can be taken as an equivalent magnetic current source, where the magnetic current flows along the slot length. To achieve maximum electromagnetic coupling between the aperture and DRA, the aperture has to be placed in a region with strongest magnetic fields. Therefore, the internal magnetic field is of great importance in the determination of the aperture location for the efficient excitation of a desired mode.

In general, the slot area has to be small enough to avoid the backward radiations. However, in many situations, the slot can be also designed as a radiator that resonates at another frequency for the realization of dual-band antenna operations.

Microstrip line feeding is another commonly used excitation method for DRAs. Figure 4.1c shows the configuration of a microstrip line–fed cylindrical DRA. The energy coupled from the microstrip line excites the horizontal magnetic dipole mode within the DRA. The amount of the coupling can be adjusted by varying the relative position of the DRA and the microstrip line. The internal magnetic fields in the DRA also determine the optimal location of the microstrip line for achieving maximum energy coupling. Microstrip line feeding is very suitable for the feeding of DRA arrays [19, 20].

4.1.2 Importance of Modal Analysis and its Challenges

As can be noted from the preceding subsection, investigations into the resonant behavior and internal fields of each radiating modes are essential for DRA designs. Specifically, exciting the modes at their resonant frequency ensures the maximum radiation efficiency. Knowledge of the internal fields is essential for determining which excitation method is best suited to excite a desired mode. Furthermore, a good knowledge of the internal fields as a function of the location is valuable in the feeding placement to allow the largest amount of power coupling between the feeding structure and the DRA. Therefore, it is important to have a good understanding of the resonant behavior and internal fields of each radiating mode in DRAs.

The numerical analysis for DRAs has been in existence for a long time. In addition to the full-wave numerical techniques that solving the Maxwell equations with external sources, source free modal analysis methods are always preferred in DRA analysis. However, modal solutions are generally only available for simple DRA shapes that are conformal to separable coordinate systems [14, 21, 22]. In general, the internal fields in DRAs are difficult to calculate, even for simple DRA shapes like the cylindrical and rectangular DRAs. Exact solutions for internal fields are only available for spherical DRAs. Many assumptions are required in the calculation of the approximate internal fields of cylindrical and rectangular DRAs [23]. Therefore, it is generally understood that it is not easy to investigate the resonant behavior and internal fields of more general or even very complicated DRA shapes.

Before the advent of our proposed characteristic mode (CM) theory for dielectric resonators, modal analysis for arbitrarily shaped DRAs was conventionally conducted by seeking the roots of the determinant of the method of moments (MoM) impedance matrix in a complex frequency plane [24, 25]. Unfortunately, the numerical implementation of this method has many serious issues. First, a significant variation of the determinant appears only in a very small region around the resonant frequency. Therefore, a very good initial guess of the root is usually required to ensure the convergence of the root searching procedure. However, a good initial guess of the root is generally unavailable [26]. Second, root seeking must be conducted in a complex frequency plane. The root seeking along the imaginary frequency axis introduces heavy computational burden. Third, the modal currents are solved from a homogeneous matrix equation. Because the right-hand side vector in the homogeneous matrix equation is a zero vector, neither the Gaussian elimination nor the more advanced solvers based on the singular value decomposition [24, 25] work well for such homogeneous matrix equation. This problem becomes more serious when the number of unknowns increases. These issues prohibit the application of the determinant root seeking method in the DRA designs. Therefore, a versatile modal analysis method is highly demanded for arbitrary DRA shapes. This is also the motivation of our contribution in the CM theory development for dielectric resonators.

4.1.3 Early Attempts to DRA Modal Analysis Using Characteristic Mode Theory

As we have discussed in the preceding chapters, the CM theory provides an elegant and useful tool to calculate the resonant frequency, modal currents, and modal fields for conducting bodies and structures embedded in multilayered medium. In the early 1970s, attempts had also been made to extend the CM theory to dielectric bodies [27, 28]. In Ref. [27], the CM formulation was developed from the volume integral equation. Although it may provide correct modal solutions for dielectric bodies, it requires large number of MoM unknowns even for an electrical small dielectric body. Therefore, solving the generalized eigenvalue equation is very time consuming. This may be the reason that the CM formulation proposed in Ref. [27] has not been widely adopted in the modal analysis of dielectric resonators.

On the other hand, the CM formulation in Ref. [28] was developed from the PMCHWT (Poggio, Miller, Chang, Harrington, Wu, and Tsai) formulation [29]. The PMCHWT formulation is a set of coupled surface integral equations (SIE). By invoking the equivalent principle for dielectric bodies, the MoM unknowns are only defined on the surface of the dielectric bodies. Therefore, it only requires a small number of unknowns to discretize the PMCHWT formulation. The computational efficiency issue in Ref. [27] is thus eliminated. Chang and Harrington [28] described two fundamental contributions for the analysis of the dielectric bodies: PMCHWT formulation and a sound idea for formulating CM equation. However, an extensive literature survey shows that the CM formulation in [28] has never been applied in the modal analysis of any dielectric resonators, although the CM formulation has been published for nearly four decades. Moreover, our careful theoretical and numerical investigations reveal that the CM formulation based on the SIE in Ref. [28] may not completely describe the natural resonant frequencies and natural modal fields of the dielectric resonators. Another independent research group at University of Ottawa, Canada, also found that the modal fields computed from the CM formulation in Ref. [28] did not provide the reasonable modal solutions for dielectric resonators [30]. They also found that the far fields of the CMs did not hold the orthogonality in the far-field zone [30]. However, orthogonality is the most important property of CMs. For convenience, we refer to the previously proposed CM formulation in Ref. [28] as the 1970s CM formulation in this book.

4.1.4 Contributions of this Chapter

In order to apply the fundamental idea of the 1970s CM formulation to the CM analysis of DRAs, this chapter aims to develop a useable CM formulation for the modal analysis of arbitrarily shaped dielectric bodies [31, 32]. Three points can be made in terms of the contribution of this chapter.

First, numerical investigation for the 1970s CM formulation is presented. Although the PMCHWT integral equation in Ref. [28] is well acknowledged in the computational electromagnetic community, the correctness of the CM formulation derived from this PMCHWT integral equation has not been validated for obtaining correct natural resonant frequencies and their corresponding modal fields yet. The careful numerical investigation in this chapter shows that the 1970s CM formulation may not be able to give the correct modal fields and resonant frequencies for dielectric resonators, even for the simplest spherical DRA.

Second, we propose a new CM formulation to complete and perfect the fundamental idea of solving CMs from the MoM matrix of the PMCHWT formulation. Based on the dependent relationship between the electric and magnetic currents, we derive a new matrix that relates the excitation term with only the electric current. Based on this new matrix, a generalized eigenvalue equation with only the electric current is developed for the CM analysis of the dielectric resonators. Characteristic electric current can be solved from this generalized eigenvalue equation. Applying the dependent relationship between the electric and magnetic currents, characteristic magnetic current can be readily found. Comparison study is performed with either experimental results or numerical results obtained in other approaches. It shows that the newly developed CM formulation for dielectric resonators can give the correct resonant frequency and characteristic fields (both internal fields and far fields).

Third, another new matrix is derived to relate the excitation term with the magnetic current. Another newly developed generalized eigenvalue equation involving only the magnetic current is then developed. It can be considered as a dual form of the previous one for predicting the resonant frequencies of DRAs. Characteristic currents can be computed in the similar manner. The two newly developed generalized eigenvalue equations can be used independently for the CM analysis of the dielectric resonators. Numerical simulations validate that they are able to provide the correct modal fields and resonant frequencies for arbitrary dielectric resonators.

In addition, as compared with the determinant root seeking method, the proposed CM formulations provide a good indicator to show the resonant frequency of each mode clearly. Moreover, the CM analysis avoids the frequency sweeping along the imaginary frequency axis. It also avoids solving the homogeneous matrix equation. These factors ensure the computational efficiency of the developed CM formulation.

4.2 CM FORMULATIONS FOR DIELECTRIC BODIES

The CM formulations are developed from the MoM impedance matrix of the PMCHWT formulation. This section begins with an introduction to the PMCHWT formulation. The generalized eigenvalue equations in terms of the electric current and magnetic current are then developed. The algorithm introduced in Chapter 2 can be applied directly to solve these newly developed generalized eigenvalue equations. Important CM quantities will be finally introduced to interpret the modal resonant properties of dielectric resonators.

4.2.1 PMCHWT Surface Integral Equations

Figure 4.2(a) shows an arbitrarily shaped homogeneous dielectric body with permittivity ε_2 and permeability μ_2. It is enclosed by its boundary surface S and is located in an infinite homogeneous medium with permittivity ε_1 and permeability μ_1. In this chapter, we assume the homogeneous dielectric body is located in free space, that is, $\varepsilon_1 = \varepsilon_0$, $\mu_1 = \mu_0$. In scattering problem, the dielectric body is illuminated by an incident plane wave $(\mathbf{E}^i, \mathbf{H}^i)$. By invoking the surface equivalence principle [33], two equivalent problems corresponding to the exterior and interior regions are formed in terms of the equivalent surface electric and magnetic current.

Figures 4.2(b) and 4.2(c) show the equivalent problems for the exterior and interior regions, respectively. In the exterior equivalent problem, the fields in the interior region are zero, whereas the fields in the exterior region are the same as the original problem. Moreover, both the interior and exterior regions are filled with the homogeneous medium (ε_1, μ_1). It allows calculating the fields using the Green's function in the homogeneous medium (ε_1, μ_1). By enforcing the continuous condition of the electric and magnetic fields on the equivalent surface, we have:

$$-\left[\mathbf{E}_1^s\left(\mathbf{J}_1\right) + \mathbf{E}_1^s\left(\mathbf{M}_1\right)\right]_{\tan} = \mathbf{E}_{\tan}^i \tag{4.1}$$

$$-\left[\mathbf{H}_1^s\left(\mathbf{J}_1\right) + \mathbf{H}_1^s\left(\mathbf{M}_1\right)\right]_{\tan} = \mathbf{H}_{\tan}^i \tag{4.2}$$

where $\mathbf{E}_1^s$ and $\mathbf{H}_1^s$ are the scattered electric and magnetic field in the external region. The subscript "tan" denotes the tangential components of the field.

In the interior equivalent problem as shown in Figure 4.2c, the fields in the exterior region are zero, whereas the fields in the interior region are the same as those in the original problem. Both of the interior and exterior regions are filled with the homogeneous medium (ε_2, μ_2). The fields can be calculated using the Green's function in the homogeneous medium (ε_2, μ_2). By enforcing the continuous condition of the electric and magnetic fields on the equivalent surface, we have another two equations:

$$-\left[\mathbf{E}_2^s\left(\mathbf{J}_2\right) + \mathbf{E}_2^s\left(\mathbf{M}_2\right)\right]_{\tan} = 0 \tag{4.3}$$

$$-\left[\mathbf{H}_2^s\left(\mathbf{J}_2\right) + \mathbf{H}_2^s\left(\mathbf{M}_2\right)\right]_{\tan} = 0 \tag{4.4}$$

where, for the sake of convenience, we use $\mathbf{E}_2^s = \mathbf{E}_2$ and $\mathbf{H}_2^s = \mathbf{H}_2$ to denote the electric and magnetic field in the interior region, respectively. As can be observed

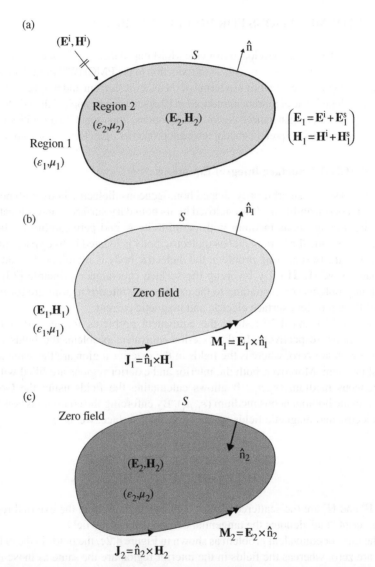

FIGURE 4.2   Surface equivalence principle of a dielectric scattering problem. (a) Original problem, (b) exterior equivalence, and (c) interior equivalence.

from Equations (4.1)–(4.4), the electric and magnetic fields in the interior and exterior regions are formulated in terms of the equivalent surface electric current and magnetic current in their respective homogeneous medium regions. In the SIEs of homogenous dielectric objects, operators L_v and K_v in the region v $(v = 1, 2)$ are defined for the convenience of field expressions:

$$L_v(\mathbf{X}) = \frac{jk_v}{4\pi}\left(\int_S \mathbf{X}(\mathbf{r}')G_v(\mathbf{r},\mathbf{r}')dS' + \frac{1}{k_v^2}\nabla\int_S \nabla'\cdot\mathbf{X}(\mathbf{r}')G_v(\mathbf{r},\mathbf{r}')dS'\right) \quad (4.5)$$

$$K_v(\mathbf{X}) = \frac{1}{4\pi} \int_S \nabla' G_v(\mathbf{r},\mathbf{r}') \times \mathbf{X}(\mathbf{r}') \, dS' \tag{4.6}$$

where $G_v(\mathbf{r},\mathbf{r}') = \dfrac{e^{-jk_v|\mathbf{r}-\mathbf{r}'|}}{|\mathbf{r}-\mathbf{r}'|}$ is the Green's function multiplied by 4π in region v and $k_v = \omega\sqrt{\varepsilon_v\mu_v}$ is the wave number in the medium (ε_v,μ_v). The electric and magnetic fields due to the electric current $\mathbf{J}_v$ and magnetic current $\mathbf{M}_v$ in the region v can be expressed using the L_v and K_v operators:

$$\mathbf{E}_v^s(\mathbf{J}_v) = -\eta_v L_v(\mathbf{J}_v) \tag{4.7}$$

$$\mathbf{E}_v^s(\mathbf{M}_v) = K_v(\mathbf{M}_v) \tag{4.8}$$

$$\mathbf{H}_v^s(\mathbf{J}_v) = K_v(\mathbf{J}_v) \tag{4.9}$$

$$\mathbf{H}_v^s(\mathbf{M}_v) = -\frac{1}{\eta_v} L_v(\mathbf{M}_v) \tag{4.10}$$

where $\eta_v = \sqrt{\mu_v/\varepsilon_v}$ is the wave impedance in the medium (ε_v,μ_v). Because $\mathbf{J}_1 = -\mathbf{J}_2$ and $\mathbf{M}_1 = -\mathbf{M}_2$, we can unify the denotations for the unknowns by introducing $\mathbf{J} = \mathbf{J}_1$ and $\mathbf{M} = \mathbf{M}_1$. $\mathbf{J}_2$ and $\mathbf{M}_2$ in Equations (4.1)–(4.4) can be further denoted as $\mathbf{J}_2 = -\mathbf{J}_1$ and $\mathbf{M}_2 = -\mathbf{M}_1$.

Therefore, there are two unknowns $\mathbf{J}$ and $\mathbf{M}$ to be determined from the four equations in (4.1)–(4.4). Various formulations can be derived by using different combinations of the four equations. The PMCHWT formulation is the most widely used one due to its accuracy, stability, and free of internal resonance property. By adding Equation (4.1) to Equation (4.3) and adding Equation (4.2) to Equation (4.4), and representing the fields using the formulations in Equations (4.7)–(4.10), we can obtain the following PMCHWT formulations:

$$\left[(\eta_1 L_1 + \eta_2 L_2)(\mathbf{J}) - (K_1 + K_2)(\mathbf{M})\right]_{\tan} = \mathbf{E}_{\tan}^i \tag{4.11a}$$

$$\left[(K_1 + K_2)(\mathbf{J}) + \left(\frac{1}{\eta_1} L_1 + \frac{1}{\eta_2} L_2\right)(\mathbf{M})\right]_{\tan} = \mathbf{H}_{\tan}^i \tag{4.11b}$$

Because the electric and magnetic currents and the electric and magnetic fields are scaled by a constant wave impedance, the PMCHWT formulations in Equations (4.11a) and (4.11b) can be transformed to the following form by multiplying η_1 to the incident magnetic field:

$$\begin{bmatrix} L_1 + \dfrac{\eta_2}{\eta_1} L_2 & -(K_1 + K_2) \\[2mm] K_1 + K_2 & L_1 + \dfrac{\eta_1}{\eta_2} L_2 \end{bmatrix} \begin{bmatrix} \eta_1 \mathbf{J} \\[2mm] \mathbf{M} \end{bmatrix} = \begin{bmatrix} \mathbf{E}_{\tan}^i \\[2mm] \eta_1 \mathbf{H}_{\tan}^i \end{bmatrix} \tag{4.12}$$

Although Equation (4.12) is equivalent to the formulations in Equations (4.11a) and (4.11b) mathematically, the MoM impedance matrix calculated from (4.12) has better matrix condition number than the one from Equations (4.11a) and (4.11b). Therefore, in the scattering problem, Equation (4.12) is often used to get fast convergence rate when iterative matrix solution algorithm is applied to solve the matrix equation. The following CM theory is also developed from Equation (4.12).

4.2.2 MoM Matrix Equation

To discretize the operator equation in Equation (4.12) into a matrix equation, the surface S is meshed into triangular elements with edge lengths of about $\lambda_0/10$, where λ_0 is the wavelength in the free space. To ensure the accuracy in high-order modes, the mesh density is suggested to increase to around $\lambda_0/20$. The equivalent currents $\mathbf{J}$ and $\mathbf{M}$ are then expanded using the same set of Rao-Wilton-Glisson (RWG) basis functions $\mathbf{f}_n(\mathbf{r})$ [34, 35]:

$$\eta_1 \mathbf{J}(\mathbf{r}) = \sum_{n=1}^{N} J_n \mathbf{f}_n(\mathbf{r}) \tag{4.13}$$

$$\mathbf{M}(\mathbf{r}) = \sum_{n=1}^{N} M_n \mathbf{f}_n(\mathbf{r}) \tag{4.14}$$

where J_n and M_n are the unknown coefficients and N is the number of unknowns. Substituting Equations (4.13) and (4.14) into the PMCHWT formulation in Equation (4.12), and testing the equations with the RWG basis function $\mathbf{f}_m(\mathbf{r})$, one can obtain a matrix equation:

$$\begin{bmatrix} \mathbf{Z}^{EJ} & \mathbf{Z}^{EM} \\ \mathbf{Z}^{HJ} & \mathbf{Z}^{HM} \end{bmatrix} \begin{bmatrix} \mathbf{J} \\ \mathbf{M} \end{bmatrix} = \begin{bmatrix} \mathbf{V}^{E} \\ \mathbf{V}^{H} \end{bmatrix} \tag{4.15}$$

where the sub-matrices $\mathbf{Z}^{EJ}$, $\mathbf{Z}^{EM}$, $\mathbf{Z}^{HJ}$, and $\mathbf{Z}^{HM}$ are $N \times N$ matrices and the MoM impedance matrix in (4.15) is a $2N \times 2N$ matrix. $\mathbf{J} = [J_n]_{N \times 1}$ and $\mathbf{M} = [M_n]_{N \times 1}$ are the vectors that consist of the coefficients for the equivalent electric and magnetic currents, respectively. $\mathbf{V}^E$ and $\mathbf{V}^H$ are the excitation vectors obtained by testing the incident electric field $\mathbf{E}^i_{tan}$ and weighted magnetic field $\eta_1 \mathbf{H}^i_{tan}$ on the equivalent surface S with $\mathbf{f}_m(\mathbf{r})$, respectively.

Based on the matrix equation in (4.15), we will develop two generalized eigenvalue equations for the CM analysis of dielectric bodies. One directly results in the characteristic electric current and another directly results in the characteristic magnetic current. To find the complete characteristic currents, dependent relationships between the electric and magnetic currents are formulated. The internal fields within the dielectric body and the radiating far fields can be computed from Equations (4.7) to (4.10) by considering the contributions of the equivalent electric and magnetic currents simultaneously.

4.2.3 Generalized Eigenvalue Equation for Characteristic Electric Current

From the PMCHWT formulation, one can readily obtain the magnetic current $\mathbf{M}$ when the electric current $\mathbf{J}$ and the external sources $(\mathbf{E}^i, \mathbf{H}^i)$ are known. In other words, $\mathbf{M}$ is dependent on $\mathbf{J}$. The explicit expression of Equation (4.15) can be written as:

$$\mathbf{Z}^{EJ} \mathbf{J} + \mathbf{Z}^{EM} \mathbf{M} = \mathbf{V}^{E} \tag{4.16}$$

$$\mathbf{Z}^{HJ} \mathbf{J} + \mathbf{Z}^{HM} \mathbf{M} = \mathbf{V}^{H} \tag{4.17}$$

Making use of (4.17), the magnetic current can be expressed as:

$$\mathbf{M} = \left(\mathbf{Z}^{HM}\right)^{-1} \left(\mathbf{V}^{H} - \mathbf{Z}^{HJ}\mathbf{J}\right) \tag{4.18}$$

Substitution of (4.18) into (4.16) and moving the excitation terms to the right-hand side yields:

$$\left[\mathbf{Z}^{EJ} - \mathbf{Z}^{EM}\left(\mathbf{Z}^{HM}\right)^{-1}\mathbf{Z}^{HJ}\right]\mathbf{J} = \mathbf{V}^{E} - \mathbf{Z}^{EM}\left(\mathbf{Z}^{HM}\right)^{-1}\mathbf{V}^{H} \tag{4.19}$$

As can be seen, Equation (4.19) includes only the electric current $\mathbf{J}$. The external source and induced electric current are thus related by a new impedance matrix:

$$\mathbf{Z}^{E} = \mathbf{Z}^{EJ} - \mathbf{Z}^{EM}\left(\mathbf{Z}^{HM}\right)^{-1}\mathbf{Z}^{HJ} \tag{4.20}$$

This new $N \times N$ matrix $\mathbf{Z}^{E}$ will facilitate the CM analysis of dielectric resonators. Different from the matrix in the Equation (4.15), which was used in the derivation of the 1970s CM formulation [28], $\mathbf{Z}^{E}$ constrains the relationship between the electric current $\mathbf{J}$ and magnetic current $\mathbf{M}$ inherently. Therefore, CM formulation derived from $\mathbf{Z}^{E}$ will produce physically reasonable CMs for dielectric bodies.

In the CM theory, the prerequisite for the characteristic electric field is that it should be equiphase on the equivalent surface S. Moreover, the equiphase characteristic electric field must lag the characteristic electric current by a constant angle (known as the characteristic angle) [36, 37]. The prerequisite rigorously defines the phase features of the characteristic electric field and the characteristic electric current on the equivalent surface S.

Different from the CM formulations developed by Garbacz [36], Harrington and Mautz proposed a straightforward method to derive the generalized eigenvalue equation for conducting bodies based on linear operator matrix [38]. The same method has also been applied to derive the generalized eigenvalue equation to solve CMs of perfectly electrically conducting (PEC) structures in multilayered medium. Following the same idea, a generalized eigenvalue equation was derived to compute the CMs of dielectric bodies using the PMCHWT formulation [28]. However, the matrix used in that derivation might not fully consider the dependent relationship between the sources $\mathbf{J}$ and $\mathbf{M}$ for the situation of source free to obtain CMs with natural mode behavior.

For the ease of understanding, we derive the generalized eigenvalue equation using $\mathbf{Z}^{E}$ by following the standard procedure proposed in Ref. [38]. In this standard procedure, a weighted eigenvalue equation is first defined for the matrix that relates the source and the field:

$$\mathbf{Z}^{E}\mathbf{J}_{n} = v_{n}\mathbf{W}\mathbf{J}_{n} \tag{4.21}$$

where v_{n} is the eigenvalue, $\mathbf{J}_{n}$ is the characteristic electric current, and $\mathbf{W}$ is a weight operator matrix to be chosen. To ensure the orthogonality of the CMs, the choice of $\mathbf{W} = \mathbf{R}^{E}$ is applied in (4.21). In addition, let $v_{n} = 1 + j\lambda_{n}$, and substituting $\mathbf{Z}^{E} = \mathbf{R}^{E} + j\mathbf{X}^{E}$ into (4.21) gives a generalized eigenvalue equation:

$$\mathbf{X}^{E}\mathbf{J}_{n} = \lambda_{n}\mathbf{R}^{E}\mathbf{J}_{n} \tag{4.22}$$

where λ_n and $\mathbf{J}_n$ are the eigenvalue and characteristic electric current. $\mathbf{R}^E$ and $\mathbf{X}^E$ are the real and imaginary Hermitian parts of $\mathbf{Z}^E$.

After the normalization of $\mathbf{J}_n$ to unit radiated power, the following orthogonal relationships are obtained:

$$\left[\mathbf{J}_n\right]^{\mathrm{T}} \cdot \mathbf{R}^E \cdot \mathbf{J}_m = \delta_{mn} \tag{4.23}$$

$$\left[\mathbf{J}_n\right]^{\mathrm{T}} \cdot \mathbf{X}^E \cdot \mathbf{J}_m = \lambda_n \delta_{mn} \tag{4.24}$$

$$\left[\mathbf{J}_n\right]^{\mathrm{T}} \cdot \mathbf{Z}^E \cdot \mathbf{J}_m = \left(1 + j\lambda_n\right)\delta_{mn} \tag{4.25}$$

where δ_{mn} is the Kronecker delta function and $\left[\cdot\right]^{\mathrm{T}}$ defines the transpose of a vector or a matrix.

Besides, the CM analysis is a source-free analysis. The CMs are independent of any external sources and excitations [36, 38]. They are only dependent on the geometry, size, and material of an electromagnetic structure. Applying the source-free condition $\mathbf{V}^H = 0$ in (4.18), one can find the resultant magnetic current $\mathbf{M}(\mathbf{J}_n)$ in terms of $\mathbf{J}_n$:

$$\mathbf{M}\left(\mathbf{J}_n\right) = -\left(\mathbf{Z}^{HM}\right)^{-1} \mathbf{Z}^{HJ} \mathbf{J}_n \tag{4.26}$$

It should be noted that characteristic electric current $\mathbf{J}_n$ works with its resultant magnetic current $\mathbf{M}(\mathbf{J}_n)$ together to produce its modal field and modal radiation pattern.

Applying the complex Poynting's theorem and the orthogonal relationships in $\mathbf{J}_n$ and $\mathbf{M}(\mathbf{J}_n)$, one can get the orthogonal relationship in the characteristic fields:

$$\frac{1}{\eta} \oiint_{S_\infty} \mathbf{E}_m^* \cdot \mathbf{E}_n \, ds = \delta_{mn} \tag{4.27}$$

4.2.4   Important CM Quantities

Owing to the orthogonality in the CMs, radiated fields and current distributions can be expanded using the CMs. In other words, the surface currents can be written as the linear combinations of the characteristic currents:

$$\mathbf{J} = \sum_{n=1}^{N} \tau_n \mathbf{J}_n \tag{4.28}$$

This formulation is known as the modal solutions of the CM theory. It states that any practically induced currents can be expanded using a complete set of orthogonal characteristic currents.

The CMs offer great flexibility to control the radiation behavior of a resonator. In light of the different current distributions, one can selectively excite a mode by carefully designing an excitation structure. As has been discussed in the preceding sections, the excitation structures can be in many different forms, including the coaxial probe feeding, aperture coupling feeding, and microstrip line feeding.

The contribution of a single mode to the total response of the system is governed by the modal significance.

Substituting Equation (4.28) into Equation (4.19) and multiplying it with $[\mathbf{J}_m]^T$, one can obtain:

$$\sum_{n=1}^{N} \tau_n [\mathbf{J}_m]^T \cdot \mathbf{Z}^E \cdot \mathbf{J}_n = [\mathbf{J}_m]^T \cdot \left[\mathbf{V}^E - \mathbf{Z}^{EM} \left(\mathbf{Z}^{HM} \right)^{-1} \mathbf{V}^H \right] \tag{4.29}$$

Making use of the orthogonality in Equation (4.25), only the term with $m = n$ exists in the left-hand side:

$$\tau_n \left(1 + j\lambda_n \right) = [\mathbf{J}_n]^T \cdot \left[\mathbf{V}^E - \mathbf{Z}^{EM} \left(\mathbf{Z}^{HM} \right)^{-1} \mathbf{V}^H \right] \tag{4.30}$$

Therefore, one can readily find the modal expansion coefficient from the following:

$$\tau_n = \frac{[\mathbf{J}_n]^T \cdot \left[\mathbf{V}^E - \mathbf{Z}^{EM} \left(\mathbf{Z}^{HM} \right)^{-1} \mathbf{V}^H \right]}{1 + j\lambda_n} \tag{4.31}$$

It shows that the contribution of each mode is scaled by a constant factor, which is known as the modal significance:

$$\mathrm{MS} = \frac{1}{\left| 1 + j\lambda_n \right|} \tag{4.32}$$

The modal significance controls the coupling of an external source ($\mathbf{V}^E$ and $\mathbf{V}^H$) to each mode. Equation (4.31) indicates that modes with large modal significances contribute more to the total response of a radiating system.

Referring to the prerequisite for CMs, another important property is the phase lag between the near fields and characteristic currents. The phase lag angle is defined as:

$$\alpha_n = 180° - \tan^{-1} \lambda_n \tag{4.33}$$

This angle is known as the characteristic angle. It is a useful indicator in some antenna designs. For example, to get optimal feeding point for circularly polarized DRAs, one must investigate the phases of the modal currents carefully.

4.2.5 Generalized Eigenvalue Equation for Characteristic Magnetic Current

The generalized eigenvalue equation in (4.22) directly computes the characteristic electric currents. The development of a dual form of Equation (4.22) that directly produces the characteristic magnetic currents is addressed in this section.

Making use of Equation (4.16), the electric current can be expressed as:

$$\mathbf{J} = \left(\mathbf{Z}^{EJ} \right)^{-1} \left(\mathbf{V}^E - \mathbf{Z}^{EM} \mathbf{M} \right) \tag{4.34}$$

Substituting Equation (4.34) into (4.17) and moving the excitation terms to the right-hand side, we obtain:

$$\left[\mathbf{Z}^{HM} - \mathbf{Z}^{HJ} \left(\mathbf{Z}^{EJ} \right)^{-1} \mathbf{Z}^{EM} \right] \mathbf{M} = \mathbf{V}^{H} - \mathbf{Z}^{HJ} \left(\mathbf{Z}^{EJ} \right)^{-1} \mathbf{V}^{E} \tag{4.35}$$

The external source and induced magnetic current are thus related by a new matrix:

$$\mathbf{Z}^{M} = \mathbf{Z}^{HM} - \mathbf{Z}^{HJ} \left(\mathbf{Z}^{EJ} \right)^{-1} \mathbf{Z}^{EM} \tag{4.36}$$

Similar to its counterpart in (4.20), the dimension size of $\mathbf{Z}^M$ is also $N \times N$.

In similar way, applying the standard procedure [38] in the new matrix $\mathbf{Z}^M$, one can obtain the dual form of the generalized eigenvalue equation in Equation (4.22):

$$\mathbf{X}^{M} \mathbf{M}_{n} = \lambda_{n} \mathbf{R}^{M} \mathbf{M}_{n} \tag{4.37}$$

where λ_n is the eigenvalue of the characteristic currents $\mathbf{M}_n$. $\mathbf{R}^M$ and $\mathbf{X}^M$ are the real and imaginary Hermitian parts of $\mathbf{Z}^M$. Both $\mathbf{R}^M$ and $\mathbf{X}^M$ are real symmetric matrices, and $\mathbf{R}^M$ is a semi-definite matrix. Therefore, all the eigenvalues and characteristic magnetic currents are real. Applying the source-free condition $\mathbf{V}^E = 0$ in (4.34), the resultant electric current $\mathbf{J}(\mathbf{M}_n)$ can be found by the following:

$$\mathbf{J}\left(\mathbf{M}_{n} \right) = -\left(\mathbf{Z}^{EJ} \right)^{-1} \mathbf{Z}^{EM} \mathbf{M}_{n} \tag{4.38}$$

By normalizing the characteristic magnetic current $\mathbf{M}_n$ to unit radiated power, the orthogonality relationships can be found:

$$\left[\mathbf{M}_{n} \right]^{\mathrm{T}} \cdot \mathbf{R}^{M} \cdot \mathbf{M}_{m} = \delta_{mn} \tag{4.39}$$

$$\left[\mathbf{M}_{n} \right]^{\mathrm{T}} \cdot \mathbf{X}^{M} \cdot \mathbf{M}_{m} = \lambda_{n} \delta_{mn} \tag{4.40}$$

$$\left[\mathbf{M}_{n} \right]^{\mathrm{T}} \cdot \mathbf{Z}^{M} \cdot \mathbf{M}_{m} = \left(1 + j\lambda_{n} \right) \delta_{mn} \tag{4.41}$$

The modal significance and characteristic angle can be also computed from Equations (4.32) to (4.33).

4.3 ANALYSIS AND DESIGN OF DRAS USING CM THEORY

In this section, the modal analyses for many commonly used DRAs are performed using the CM theory we have developed in the preceding sections. In these examples, the resonant frequency of each radiating mode is identified through the observation of the modal significance peaks over a frequency band. To further investigate the resonant behavior and radiation performance of each mode, the internal fields and far-field radiation pattern of each CM are plotted. For convenience, we referred to the internal electric and magnetic fields in the near-field region as the modal electric field and modal magnetic field, respectively. The far-field radiation pattern due to the

TABLE 4.1 The Analyzed DRAs and Corresponding Excitation Methods

DRAs	Excitation methods
Cylindrical DRA	Coaxial probe
	Aperture coupling
Spherical DRA	Coaxial probe
	Aperture coupling
Rectangular DRA	Aperture coupling
Triangular DRA	Coaxial probe
Notched rectangular DRA	Coaxial probe

modal currents is called as the modal radiation pattern. The modal fields clearly illustrate the variation of the electric and magnetic fields with the locations inside the DRA. They are helpful to determine which type of excitation method is better to excite a desired mode. They also show the optimal location of the excitation structure for achieving maximum energy coupling between the excitation structure and the desired mode of a DRA.

In the design stage, we implement either the aperture coupling excitation or coaxial probe excitation to the DRAs. The internal fields and radiation pattern due to the excitation structures are compared with the modal fields and modal radiation patterns obtained from the CM analysis. The very good agreements demonstrate the following three important points:

1. The CM theory for dielectric resonators provides the correct modal solutions (including the resonant frequencies, characteristic currents, modal fields, and modal radiation patterns) for DRAs.
2. The modal solutions solved from the developed CM theory can be physically excited.
3. The modal solutions clearly illustrate how to select excitation method and how to do feeding placement.

Table 4.1 gives an overview of the various DRAs to be analyzed using the newly proposed CM theory. The excitation methods for each of the DRA are also listed.

4.3.1 Cylindrical Dielectric Resonator Antennas

4.3.1.1 Characteristic Mode Analysis and Discussions The first comprehensive study to examine the radiation characteristic of a cylindrical DRA is reported in Ref. [3]. In general, the resonant modes of an isolated cylindrical DRA can be categorized into the following three types: TE mode, TM mode, and hybrid modes. This modal nomenclature comes after the way defined in Ref. [39]. It is later widely adopted in the DRA research area. Among various modes in the cylindrical DRAs, the $TE_{01\delta}$ mode, $TM_{01\delta}$ mode, and $HEM_{11\delta}$ mode (hybrid mode with dominant E_z component) are the most widely used modes for radiation purpose. The $TE_{01\delta}$ mode, $TM_{01\delta}$ mode, and $HEM_{11\delta}$ mode radiate similarly to the short vertical magnetic dipole, short vertical electric dipole, and short horizontal magnetic dipole, respectively. The subscript numbers under the mode name reflect the variation of the fields in the azimuth, radial, and axial directions, respectively. The δ value varies in the range between zero and one. It indicates the resonant behavior along the z-axis of a dielectric cylinder. The resonant frequency of the $\text{TE}_{01\delta}$ mode, $\text{TM}_{01\delta}$ mode, and $\text{HEM}_{11\delta}$ mode can be obtained by solving the wave number k_0 from the following Equations [40]:

$$k_0 a = \frac{2.327}{\sqrt{\varepsilon_r + 1}} \left(1 + 0.2123 \frac{a}{h} - 0.00898 \left(\frac{a}{h} \right)^2 \right), \text{ for TE}_{01\delta} \text{ mode} \qquad (4.42)$$

$$k_0 a = \frac{\sqrt{3.83^2 + \left(\pi a / 2h \right)^2}}{\sqrt{\varepsilon_r + 2}}, \text{ for TM}_{01\delta} \text{ mode} \qquad (4.43)$$

$$k_0 a = \frac{6.324}{\sqrt{\varepsilon_r + 2}} \left(0.27 + 0.36 \frac{a}{2h} + 0.02 \left(\frac{a}{2h} \right)^2 \right), \text{ for HEM}_{11\delta} \text{ mode} \qquad (4.44)$$

where ε_r is the dielectric constant of the DRA, and a and h are the radius and height of the dielectric cylinder, respectively. Exact formulations for resonant frequencies of much higher order modes are not available, and there is no exact solution exists to calculate the internal fields and radiation patterns of these resonant modes. An approximation solution to the field pattern of cylindrical DRAs was achieved by introducing magnetic wall boundary condition on the surfaces parallel to the z-axis [41]. This approach was shown to be valid for DRAs with high dielectric constant. Its versatility in DRAs with low dielectric constant has to be examined carefully.

As can be seen, although the cylindrical DRA has very simple geometry, there are many difficulties in solving resonant frequency and modal fields. Therefore, we are going to perform the modal analysis of an isolated cylindrical dielectric resonator using the developed CM theory. The radius and height of the isolated cylindrical dielectric resonator we consider in this section are $a = 5.25$ mm and $2h = 4.6$ mm, respectively. The dielectric constant is $\varepsilon_r = 38$. Figure 4.3a shows the modal significance computed from the CM formulations derived from the matrix Z^E and Z^M, respectively. As expected, the obtained resonant frequencies agree well with each other.

To provide a full picture of the resonant behaviors of much higher order modes, Figures 4.3b and 4.3c plot the modal significances and absolute value of the eigenvalues for the first 100 modes, respectively. The color of each line is defined according to the increase of the mode's order. Lines with dark color represent the lower order modes, and lines with light color represent the higher order modes. As can be seen, the modal significance decreases from one to nearly zero as the mode index increases. Higher order modes never have large modal significances in the lower order modes' resonant frequency bands. Figure 4.3c shows that the eigenvalues of the lower order resonant modes have absolute values less than 1.0. Both the modal significance and eigenvalues are good indicators to identify the resonant frequency of each mode. It can be observed that up to a frequency of 7.75 GHz, the cylindrical resonator resonates at eight frequencies. This conclusion is the same as that was stated in Ref. [41]. Therefore, the CM formulations in this chapter have the capability to find all the

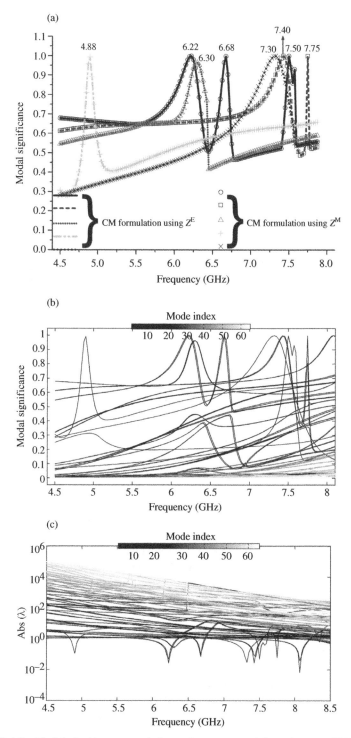

FIGURE 4.3 Modal significances and eigenvalues computed from the new CM formulations. Modal significances computed from the (a) dual CM formulations and (b) the first 100 modes, and (c) absolute value of the eigenvalues for the first 100 modes.

TABLE 4.2   Resonant Frequencies of the Cylindrical DRA

Mode	Resonant frequencies, GHz		
	Characteristic mode	Determinant root seeking method [25]	Measured results [24]
$TE_{01\delta}$	4.88	4.83	4.85
$HEM_{11\delta}$	6.30	6.33	—
$HEM_{12\delta}$	6.68	6.63	6.64
$TM_{01\delta}$	7.50	7.52	7.60
$HEM_{21\delta}$	7.75	7.75	7.81

Radius of the cylindrical resonator $= 5.25$ mm; height $= 4.6$ mm; $\varepsilon_r = 38$.

resonant modes over a given frequency band. However, the determinant root seeking method only found five resonant modes in the same frequency band [25]. Table 4.2 compares the resonant frequencies with those obtained from the determinant root seeking method and the measurement [24, 25]. It is observed that the results are very close to each other. The errors of the CM results to the measured results are less than 1.3%.

Moreover, a classification of these resonant modes requires the calculations of the modal fields. Table 4.3 displays the modal fields and modal radiation patterns of all of the eight modes, respectively. Again, as compared with the modal fields of the $TE_{01\delta}$ mode, $HEM_{11\delta}$ mode, $HEM_{12\delta}$ mode, $TM_{01\delta}$ mode, and $HEM_{21\delta}$ mode in Ref. [25], the agreement is very good. Modal fields of the other three modes were not given in Ref. [24]. The modal fields and modal radiation patterns computed from Equation (4.37) are the same as those computed from Equation (4.22).

To numerically investigate the 1970s CM formulation, Figure 4.4 shows the modal significances and eigenvalues solved from the 1970s CM formulation. The curves are also assigned with colors to identify the mode orders. As can be seen, there are a large number of modes having modal significances close to one and eigenvalues less than 1.0. In addition, the peaks of the modal significances and the nulls in the absolute value of the eigenvalues have not located correctly at the expected resonate frequencies. According to the definitions for the eigenvalue and modal significance, Figure 4.4 illustrates that there should be a large number of resonant modes existing in the 4.5–8 GHz band. However, this is not the truth and the observation goes against the true physics of the cylindrical resonator. The modal fields computed from the 1970s CM formulation also lack meaningful physics. These non-reasonable modes may be due to the ignorance of the relationships between the electric current $\mathbf{J}$ and magnetic current $\mathbf{M}$.

4.3.1.2 Excitation of $HEM_{11\delta}$ Mode Using Coaxial Probe

From Table 4.3, we have observed that the CMs of a cylindrical DRA feature with a variety of modal radiation patterns. In particular, the modal radiation pattern of the $HEM_{11\delta}$ mode is a standard broadside radiation pattern, which is very popular in many applications. Previous studies have shown that coaxial probe feeding can be introduced to excite

TABLE 4.3 Modal Fields and Modal Radiation Patterns of the Cylindrical DRA

	Modal field (electric field)	Modal radiation pattern
TE$_{01\delta}$, 4.88 GHz		
TE$_{10\delta}$, 6.22 GHz		
HEM$_{11\delta}$, 6.30 GHz		
HEM$_{12\delta}$, 6.68 GHz		
Dominant mode at 7.30 GHz		
Dominant mode at 7.40 GHz		

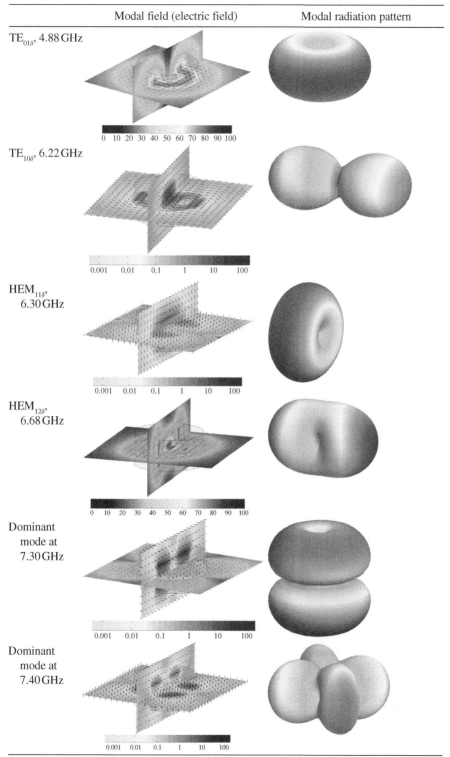

(Continued)

TABLE 4.3   (*Continued*)

	Modal field (electric field)	Modal radiation pattern
$TM_{01\delta}$, 7.50 GHz		
$HEM_{21\delta}$, 7.75 GHz		

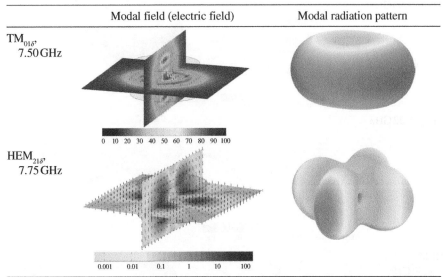

the $HEM_{11\delta}$ mode [3, 15]. One of the most important conclusions from these early studies was that the probe has to be positioned away from the cylindrical DRA's axis and near the wall of the DRA. This conclusion came from the parameter studies in the probe locations. When the probe is positioned in the aforementioned location, the DRA achieves the best input impedance matching.

On the other hand, the modal fields of the $HEM_{11\delta}$ shows that there are strong electric fields in the aforementioned location. Meanwhile, the coaxial probe is equivalent to a vertical electric current source. To achieve the largest amount of energy coupling, the coaxial probe has to be placed in the region having strong electric field. This is the reason that a displaced probe should be employed for achieving broadside radiation pattern. Figure 4.5 shows a typical configuration of a probe-fed cylindrical DRA for the excitation of the $HEM_{11\delta}$ mode [15]. With this feeding configuration, broadside radiation pattern can be obtained and the input impedance has zero reactance components at the resonant frequency of the $HEM_{11\delta}$ mode [15].

4.3.1.3   Excitation of $TE_{01\delta}$ Mode Using Aperture Coupling   As can be observed from the modal fields and modal radiation pattern of the $TE_{01\delta}$ mode, a half-split cylindrical DRA placed on a PEC ground plane can be also excited to get broadside radiation pattern. Figure 4.6 shows a half-split cylindrical DRA along with the aperture coupling structure [18]. The $TE_{01\delta}$ mode is excited by using the aperture coupling feeding. Because the strongest magnetic fields appear at the center of the cylindrical DRA, and the aperture coupling is equivalent to a magnetic current source, the best position of the slot should be at the center of the cylindrical DRA. Figure 4.7 shows the internal electric field and radiation pattern of the aperture coupling–fed half-split cylindrical DRA. Although we have introduced the ground

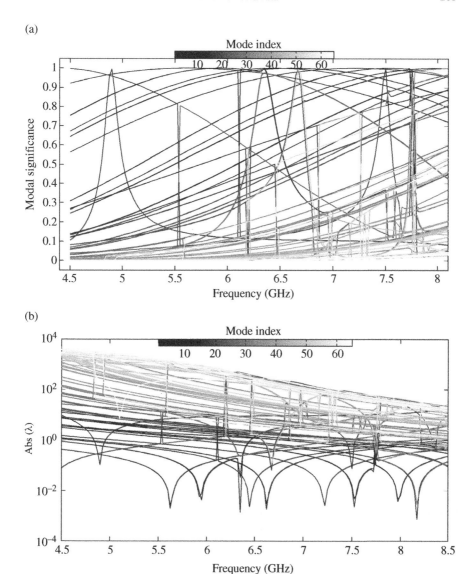

FIGURE 4.4 Modal significances and eigenvalues computed from the 1970s CM formulation. (a) Modal significances of the first 100 modes and (b) absolute value of the eigenvalues for the first 100 modes.

plane and the excitation structure, the excited electric field and radiation pattern are good approximations to the modal fields and modal radiation pattern of the $TE_{01\delta}$ mode. This example again shows that the CMs provide clear guidelines for the feeding design of a particular mode. In turn, the excited fields (near fields and far fields) demonstrate that the CMs can be practically excited by designing the excitation structures properly.

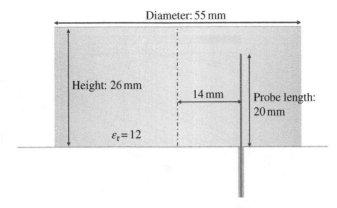

FIGURE 4.5 The configuration of a probe-fed cylindrical DRA for the excitation of $HEM_{11\delta}$ mode.

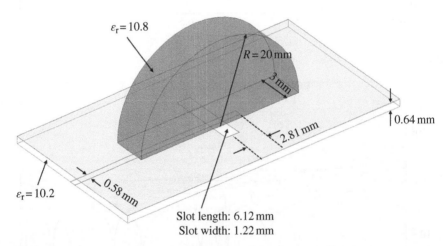

FIGURE 4.6 The configuration of an aperture coupling–fed cylindrical DRA for the excitation of $TE_{01\delta}$ mode.

4.3.2 Spherical Dielectric Resonator Antennas

4.3.2.1 Characteristic Mode Analysis and Discussions As the second example, CM analysis for a spherical dielectric resonator of radius 12.5 mm and dielectric constant $\varepsilon_r = 9.8$ is presented. Figure 4.8 shows the modal significance in the 3–6 GHz band. Again, from the peaks of the modal significance, one can observe the first three resonances appear at 3.67, 5.12, and 5.30 GHz. As can be recognized from the modal fields in Table 4.4, the resonant modes are actually the TE_{111} mode, TM_{101} mode, and TE_{221} mode of the spherical dielectric resonator. The TM_{101} mode's resonant frequency is approximately 40% higher than that of the TE_{111} mode. This observation agrees with the empirical results presented in Ref. [14]. Moreover, as can be observed from Table 4.4, the TE_{111} mode radiates similarly to a magnetic dipole,

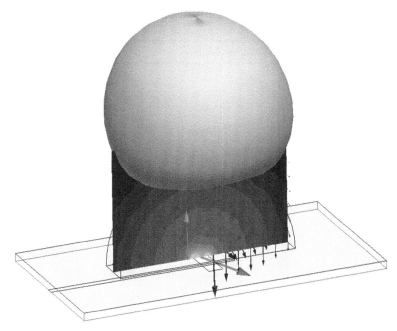

FIGURE 4.7 The internal electric field and radiation pattern due to the aperture coupling feeding.

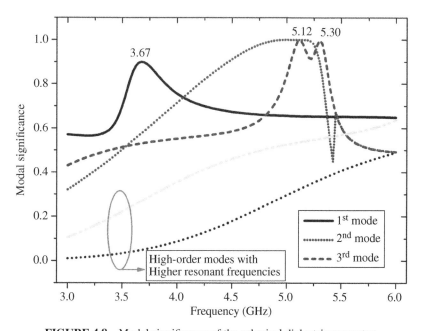

FIGURE 4.8 Modal significance of the spherical dielectric resonator.

TABLE 4.4   Modal Fields and Modal Radiation Patterns of An Isolated Spherical DRA

	Modal field (electric field)	Modal radiation pattern
TE_{111}, 3.67 GHz		
TM_{101}, 5.12 GHz		
TE_{221}, 5.30 GHz		

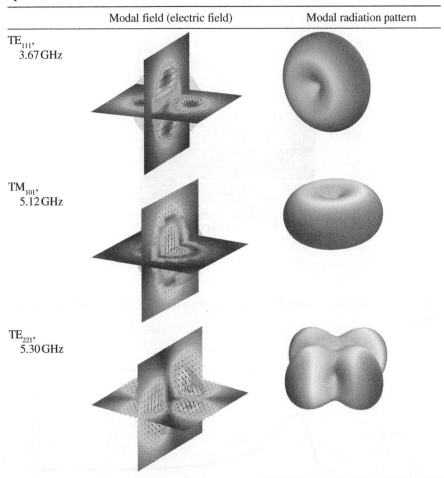

TABLE 4.5   Resonant Frequencies of the Spherical DRA

	Resonant frequencies, GHz		
Mode	Characteristic mode	MoM [14]	Measured results [14]
TE_{111}	3.67	3.500	3.7
TM_{101}	5.12	5.245	—
TE_{221}	5.30	5.267	5.3

Radius of the spherical resonator = 12.5 mm; $\varepsilon_r = 9.8$.

whereas the TM_{101} mode radiates similarly to an electric monopole. From the modal radiation patterns, one can observe that the TE_{111} mode is a broadside mode, and the TM_{101} mode and the TE_{221} mode are end-fire modes.

As expected, the TM_{101} mode and TE_{221} mode mix in a narrow frequency band [14]. Besides, Figure 4.8 also shows the modal significances for another two higher-order modes. It can easily infer that they will have their peaks at much higher frequencies. Table 4.5 compares the resonant frequencies with the MoM and measurement results [14]. It shows that the CM results agree well with the measurement. The error to the measurement results is less than 0.8%.

4.3.2.2 Excitation of TE_{111} Mode and TM_{101} Mode Using Coaxial Probe

The modal fields solved from CM theory provide helpful guidelines to the feeding designs for the excitation of the broadside TE_{111} mode and end-fire TM_{101} mode. First, because the minimum electric field intensity of the TE_{111} mode is at the center, one cannot excite this mode when the coaxial probe is near the center of the sphere. Therefore, to get the broadside radiation pattern, the probe should be placed at the location having maximum electric field intensity. This is the reason that a displaced probe away from the center is required for the excitation of the broadside TE_{111} mode in spherical DRAs.

Second, to achieve end-fire radiation from a spherical DRA, one usually places the probe at the center of the sphere. This is because the strong electric field of the TM_{101} mode is at the center. It should also be noted that there are overlapped regions having strong electric fields in the TE_{111} and TM_{101} modes. It indicates that both the modes can be excited by placing probe in this overlapped region. Figure 4.9 shows the configuration of the probe-fed spherical DRA. The reflection coefficient of this probe-fed DRA is shown in Figure 4.10. As can be seen, although the probe has not been placed at the center of the DRA for the TM_{101} mode, the TE_{111} mode at 3.6 GHz, and the TM_{101} mode at 5.2 GHz can be excited simultaneously with the same coaxial probe.

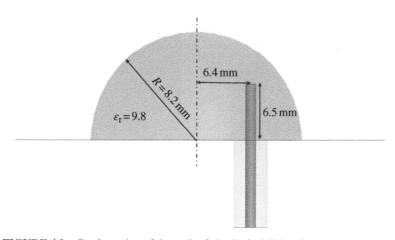

FIGURE 4.9 Configuration of the probe-fed spherical dielectric resonator antenna.

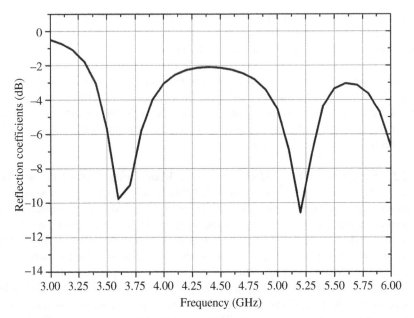

FIGURE 4.10   Reflection coefficient of the probe-fed spherical dielectric resonator antenna.

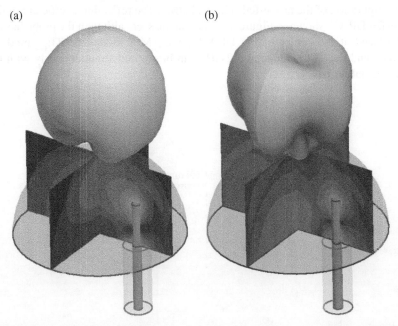

FIGURE 4.11   Radiation pattern and internal electric fields due to the coaxial probe feeding.
(a) 3.6 GHz and (b) 5.2 GHz.

Figure 4.11 shows the radiation pattern and internal electric fields due to the coaxial probe feeding in Figure 4.9. As can be expected, broadside radiation pattern is obtained at 3.6 GHz, and monopole-like radiation pattern is obtained at 5.2 GHz. Moreover, the maximum internal electric fields at 3.6 GHz appear at the upper half of vertical plane, whereas the maximum internal electric fields at 5.2 GHz appear at the center of the dielectric sphere. These internal electric field distributions approximate well to the modal fields of the TE_{111} mode and TM_{101} mode. In summary, this example illustrates the following:

- The developed CM theory can accurately determine the resonant frequency of the spherical DRA.
- The modal fields explicitly give the instructions of how to place the coaxial probe.
- The CMs solved from the present CM theory can be physically excited.

4.3.2.3 Excitation of TE_{111} Mode Using Aperture Coupling In the last section, we have demonstrated the excitation of the TE_{111} mode using coaxial probe feeding. Alternatively, it is also possible to excite the TE_{111} mode using the aperture coupling feeding [42]. Because the aperture coupling feeding is equivalent to a magnetic current source, it has to be placed at the region with the strongest magnetic fields. Figure 4.12 shows the modal magnetic fields for the TE_{111} mode of an isolated spherical DRA. As can be seen, the strongest magnetic field is at the center of the dielectric sphere. Therefore, the largest amount of energy coupling can be achieved when the slot is placed at the center of the spherical DRA. This is in contrast to the probe fed case, where a center probe feeding will fail to excite the TE_{111} mode.

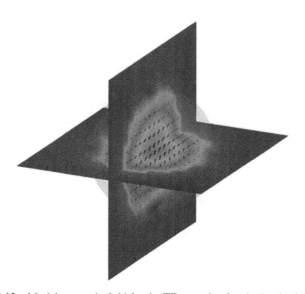

FIGURE 4.12 Modal magnetic field for the TE_{111} mode of an isolated spherical DRA.

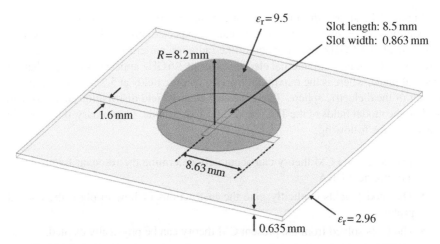

FIGURE 4.13   Configuration of an aperture coupling–fed hemi-spherical DRA.

Figure 4.13 shows the configuration of an aperture coupling fed–hemi-spherical DRA. The corresponding full-wave simulated reflection coefficient, the internal magnetic fields, and the radiation pattern at the resonant frequency 5.5 GHz are shown in Figure 4.14. With the aperture coupling feeding, a broadside radiation pattern is obtained. Moreover, the internal magnetic field distribution shows that the strongest magnetic field takes place at the center of the dielectric sphere. It is a good approximation to the modal magnetic field as shown in Figure 4.12. This example demonstrates that the modal magnetic fields in DRA are helpful in the aperture coupling feeding placement.

4.3.3 Rectangular Dielectric Resonator Antennas

As introduced in Chapter 1, the dielectric waveguide model [43] is often employed to predict the resonant frequency and the Q factor of isolated rectangular DRAs. Based on the prior knowledge of the lowest order mode, the internal fields of the lowest TE_{111}^z mode in a rectangular DRA were derived from the z-directed magnetic potential [44]. For higher order modes that cannot be simplified into the simple "magnetic dipole mode" and "electric dipole mode," it is not easy to get the internal fields in rectangular DRAs. As we shall see, the CM theory provides an efficient approach to calculate the resonant frequencies, modal fields, and modal radiation patterns for the fundamental modes and higher order modes. Examples are also presented to demonstrate that the CMs can be physically excited by designing the feeding structures following the useful information in the modal fields.

4.3.3.1 Characteristic Mode Analysis   Figure 4.15 shows the geometry of an aperture coupling–fed rectangular DRA. Detailed dimensions of the DRA and the aperture coupling structure are illustrated. The CM analysis is performed for the DRA without considering the aperture, the 0.5 mm thick dielectric substrate, and the microstrip feeding line. Figure 4.16 shows the modal significance over the frequency band of 4–7 GHz. As can be seen, the modal significance is close to one in

(a)

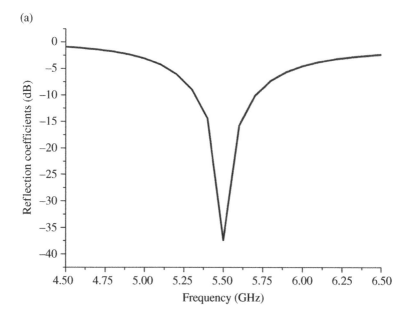

(b)

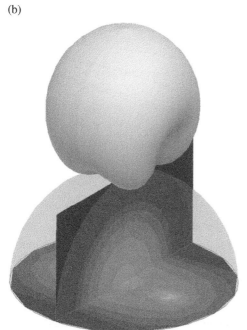

FIGURE 4.14 Simulation results of the aperture coupling–fed hemi-spherical DRA. (a) The reflection coefficient and (b) the internal magnetic fields and radiation pattern at 5.5 GHz.

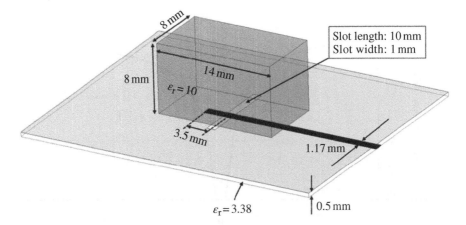

FIGURE 4.15   Geometry of the aperture coupling–fed rectangular DRA.

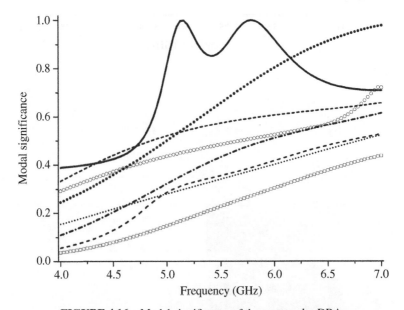

FIGURE 4.16   Modal significance of the rectangular DRA.

the frequency band of 5–6 GHz. It illustrates that the lowest order mode resonates in this frequency band. In order to investigate the resonant behavior of the rectangular DRA, the modal magnetic fields and modal radiation patterns for the first three lowest order modes are plotted in Figure 4.17. All of the modal fields and modal radiation patterns are calculated at 5.8 GHz. The field distributions and the corresponding radiation patterns are stable over the resonant frequency band. As can be observed, the first three modes have different resonant behaviors. In Figure 4.17a, both the modal magnetic field and the modal radiation pattern show that the first mode radiates like a horizontal magnetic dipole. It agrees with the modal analysis

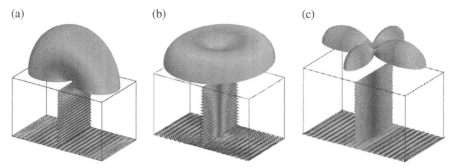

FIGURE 4.17   Modal magnetic fields and modal radiation patterns of the rectangular DRA at 5.8 GHz. (a) The first mode, (b) the second mode, and (c) the third mode.

results in Ref. [44]. The magnetic field is parallel to the width of the rectangular DRA, with its strongest magnetic field intensity locating at the center of the bottom surface of the rectangular DRA.

In order to excite the first mode for achieving broadside radiation pattern using the aperture coupling feeding, the modal magnetic field shown in Figure 4.17a gives two important points for the design of the aperture coupling feeding:

1. The length of the slot (aperture on the ground plane) has to be parallel to the width of the rectangular DRA. With this orientation, the magnetic field polarization in the slot is consistent with the modal magnetic field. This is the first prerequisite to ensure large power coupling.
2. The slot should be placed at the center of the bottom surface of the rectangular DRA. In this region, there is the strongest modal magnetic field intensity. This is the second prerequisite to ensure the large power coupling.

Considering these two points simultaneously, the resultant aperture coupling excitation structure as shown in Figure 4.15 is obtained.

The second- and third-order modes are end-fire radiation modes. In particular, the second-order mode has a monopole-like radiation pattern. The modal magnetic field illustrates that a magnetic current loop located on the bottom surface of the rectangular DRA will excite the second-order mode. In the third mode, there are many severe lobes in the radiation pattern. People are usually not interested in such kinds of radiation patterns. Except for the low modal significance, this is another reason that high-order modes are usually not excited in practical antenna designs.

4.3.3.2   Excitation of Dominant Mode Using Aperture Coupling   As we have discussed in the previous section, the aperture coupling feeding as shown in Figure 4.15 is adopted to excite the dominant mode in the band around 5.8 GHz. Full-wave simulation is conducted to examine the radiation performance of the aperture coupling–fed rectangular DRA. Figure 4.18 shows the reflection coefficient of the aperture coupling–fed rectangular DRA. It clearly shows the resonance around

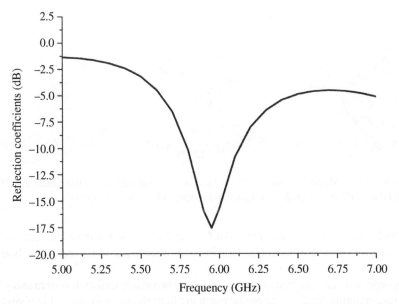

FIGURE 4.18 Simulated reflection coefficient of the aperture coupling–fed rectangular DRA.

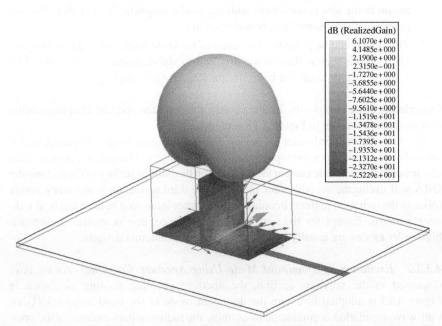

FIGURE 4.19 Simulated internal magnetic fields and radiation pattern of the aperture coupling–fed rectangular DRA at 5.95 GHz.

5.8 GHz. Far-field radiation patterns and internal magnetic fields due to the aperture coupling excitation are shown in Figure 4.19. As can be seen, the radiation pattern approximates well to the modal radiation pattern as shown in Figure 4.17a. Both have wider beam width in the elevation plane that is parallel to the length of the rectangular DRA. Moreover, the excited internal magnetic field is also similar to the modal magnetic fields in Figure 4.17a. It is clear that the slot in the ground plane works as a horizontal magnetic dipole, which agrees with the equivalent magnetic current source in the aperture coupling feeding.

4.3.4 Triangular Dielectric Resonator Antennas

The triangular DRA was introduced as a promising candidate for low-profile and compact DRA designs [45, 46]. As compared with the traditional cylindrical and rectangular DRAs, triangular DRA with the same size has lower resonant frequencies. In other words, for a given operating frequency, triangular DRA has more compact size than rectangular and cylindrical DRAs. The resonant frequencies of an equilateral triangular DRA can be solved using a simple waveguide model with magnetic walls [47]. By enforcing the continuity of the tangential field on the top and bottom surfaces of the equilateral triangular DRA, the resonant frequency can be solved from the resultant transcendental equation. This approach is best suited to triangular DRAs with very high dielectric constant. However, it is not easy to compute the internal fields in triangular DRAs, especially those for high-order modes.

4.3.4.1 Characteristic Mode Analysis and Discussion   The modal analysis for triangular dielectric resonator is performed using the CM theory for dielectric resonators. Figure 4.20 shows the geometry of an equilateral triangular DRA with a coaxial probe feeding. Again, the coaxial cable and the extended probe are not included in the CM analysis. Figure 4.21 shows the modal significance of the equilateral triangular DRA in the 1.2–1.8 GHz band. As can be observed from the peaks of the modal significance, there is a dominant resonant mode in the frequency band of 1.4–1.6 GHz. Figure 4.22 shows the modal fields and modal radiation patterns for the first four modes at 1.5 GHz. The first two modes can be classified as the broadside radiation

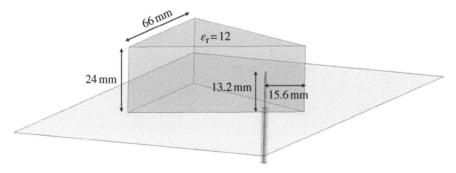

FIGURE 4.20 Geometry of the probe-fed triangular DRA.

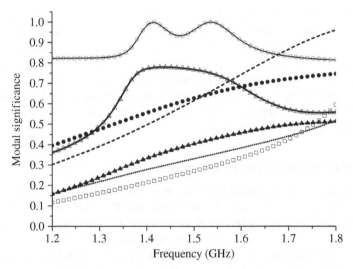

FIGURE 4.21 Modal significance of the triangular DRA.

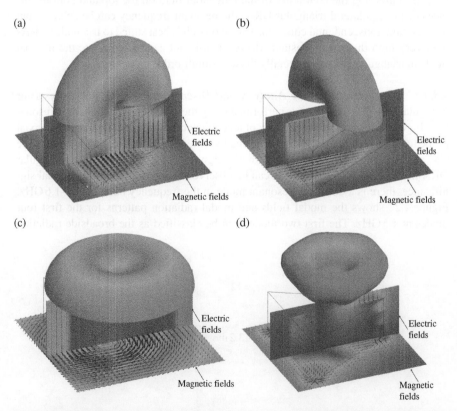

FIGURE 4.22 Modal fields and modal radiation patterns of the triangular DRA at 1.5 GHz. (a) The first mode, (b) the second mode, (c) the third mode, and (d) the fourth mode.

modes, whereas the third and fourth modes are end-fire radiation modes. We can also note that although the first two modes have broadside modal radiation patterns, the radiation mechanisms are quite different. The polarization properties of these two modal radiation fields also differ with each other. In Figure 4.22a, the internal electric fields show that a coaxial probe can be placed in the elevation plane to excite the first mode. The location of the probe can be determined by observing the maximum electric field in the elevation plane. Alternatively, Figure 4.22b shows that one can achieve broadside radiation pattern by exciting the second mode using the aperture coupling method. The slot in the ground plane has to be parallel with the length of the elevation plane as shown in Figure 4.22b. Meanwhile, the slot should be positioned in the region near the edge of the triangle. With this configuration, the modal magnetic field achieves the largest amount of power coupling from the slot.

Figure 4.22c shows that the third mode has an omnidirectional modal radiation pattern. In order to excite this mode, the modal magnetic field shows that a magnetic current loop can be introduced to the bottom surface of the triangular DRA. The center of the loop should be near the edge of the triangle. The fourth mode as shown in Figure 4.22d has a monopole-like modal radiation pattern. The modal fields show that the strongest electric and magnetic field scatters in many small regions within the DRA. Therefore, it is not easy to excite this mode simply by using a single electric current source or magnetic current source.

4.3.4.2 Excite the Dominant Mode Using Coaxial Probe Feeding

As have been discussed in the last section, coaxial probe feeding can be employed to excite the first mode for broadside radiation pattern. The modal fields in Figure 4.22a explicitly

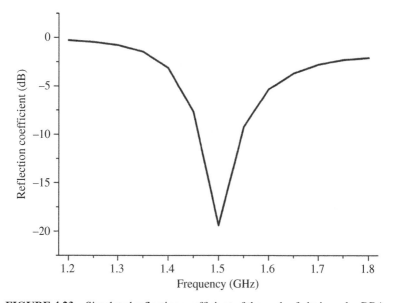

FIGURE 4.23 Simulated reflection coefficient of the probe-fed triangular DRA.

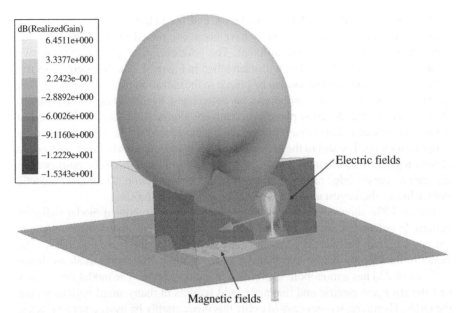

FIGURE 4.24 Simulated internal fields and radiation pattern of the probe-fed triangular DRA at 1.5 GHz.

illustrate the best location of the probe as shown in Figure 4.20. Figure 4.23 shows the full-wave simulated reflection coefficient of the probe-fed equilateral triangular DRA. It clearly shows that the DRA achieves very good input impedance matching in the frequency band centered at 1.5 GHz, which agrees well with the resonant frequency band predicted by the CM analysis.

Figure 4.24 shows the internal field and far-field radiation pattern due to the coaxial probe excitation. It can be observed that the radiation pattern is similar to the modal radiation pattern given in Figure 4.22a. Both of them have wider beam width in the internal electric field plane. Moreover, the internal field also approximates to the modal field. The regions that have the largest field intensity are consistent with each other.

4.3.5 Notched Rectangular Dielectric Resonator Antenna

To further illustrate the excitations of CMs in practical DRA design, we consider a notched rectangular dielectric resonator. Figure 4.25 shows the geometry of the notched rectangular DRA reported in [48]. The dielectric constant is $\varepsilon_r = 9.4$. Figure 4.26 shows the modal significances of the notched rectangular dielectric resonator. The peaks in the modal significance show that there are two resonant modes in the 2.42–3.50 GHz band. The first mode resonates at 2.42 GHz and the second one resonates at 2.93 GHz. Moreover, it can be observed that all the three modes mix beyond 3.5 GHz. It usually leads to the high cross polarization in the radiation pattern.

In Ref. [48], a strip line was introduced on the narrow wall to excite the notched DRA. Figure 4.26 also gives the simulated and measured reflection coefficients. The

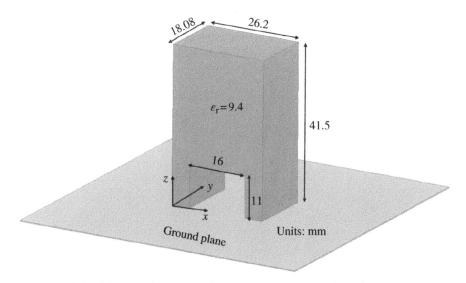

FIGURE 4.25 Configurations of the notched rectangular dielectric resonator.

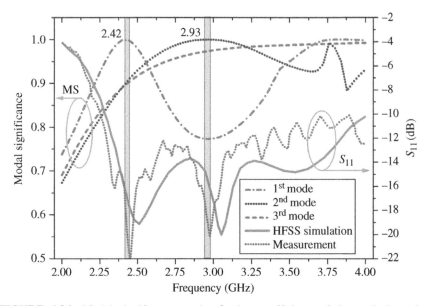

FIGURE 4.26 Modal significances and reflection coefficients of the excited notched rectangular dielectric resonator.

measurement results show that the two modes resonant at 2.46 and 2.99 GHz, respectively. They are very close to the CM analysis results. The errors of the CM analysis results to the measurement results are less than 2%. Moreover, from the frequency band that the modal significances of the first two modes are large enough, we observe very good input impedance matching. Therefore, the modal significance is also a good indicator to predict the operation bandwidth of a resonator.

Figure 4.27 shows the modal fields and modal radiation patterns in the *xoz* plane. Both the resonant modes have broadside modal radiation patterns. Therefore, the two

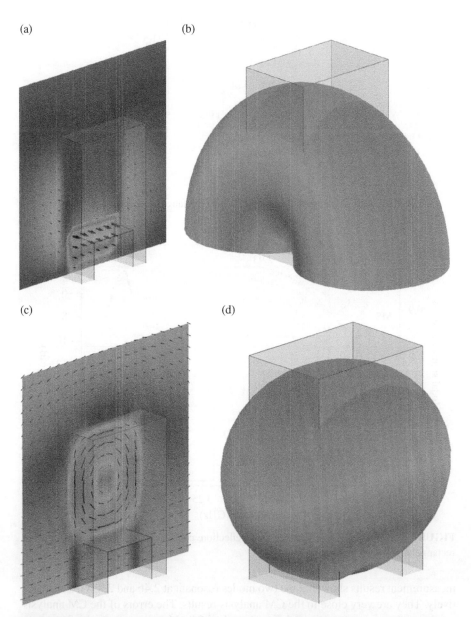

(a)   (b)

(c)   (d)

FIGURE 4.27 Characteristic modes of the notched rectangular DRA. (a) Modal electric fields at 2.42 GHz, (b) modal radiation pattern at 2.42 GHz, (c) modal electric fields at 2.93 GHz, and (d) modal radiation pattern at 2.93 GHz.

modes can be excited and merged together to increase the bandwidth. In summary, the notched rectangular DRA example shows that closely spaced CMs with similar modal radiation patterns can be physically excited and combined for wideband applications. It also demonstrates the applicability of the CM formulations for dielectric resonators with very complicated geometry.

4.4   COMPUTATIONAL EFFICIENCY

All these examples are simulated on a personal computer with a 3.4 GHz Intel Core and 8 GB RAM. Table 4.6 gives the CPU time of the CM analysis for the cylindrical DRA, spherical DRA, and notched rectangular DRA at each frequency. The CPU time includes the time for the computation of the reduced matrix in Equations (4.20) and (4.36), solving the generalized eigenvalue equation in Equations (4.22) or (4.37), and the modal tracking procedure at each frequency. It shows that the CPU time for the largest problem is less than 30 s. It is acceptable for such electrically small and medium problems (MoM unknowns ≈ 2000). Therefore, the computational efficiency for such problems is good enough for DRA design using a common personal computer.

As discussed in the literature [38, 49, 50], the developments of the CM theory (either for the conventional PEC problem [38] or the dielectric problem discussed here) are intended for the analysis of electromagnetic problems with electrically small and medium sizes. In these problems, only a few modes are needed to characterize the radiation and scattering properties. However, for electrically large problems, there are too many closely spaced resonances to identify using the CM theory. Therefore, the CM theory is not so popular in the electrically large problems. In principle, CM analysis for the electrically large problems could be implemented using high-performance computing techniques through solving the generalized eigenvalue equations proposed in this book.

4.5   CONCLUSIONS

This chapter addresses the modal analysis of dielectric resonators using the CM theory. Based on the PMCHWT formulation, two generalized eigenvalue equations are proposed for the CM analysis of dielectric resonators. CM analysis for a variety of dielectric resonators is presented. Numerical results show that the CM theory is able

TABLE 4.6   CPU Time for the CM Analysis of Three Types of DRAs

	Cylindrical DRA (Section 4.3.1.1)	Spherical DRA (Section 4.3.2.1)	Notched rectangular DRA (Section 4.3.5)
Number of triangle elements	206	446	714
Number of unknowns	618	1338	2142
Number of modes	15	25	10
CPU time (s)	2.59	11.83	29.84

to predict the resonant frequency for each radiating mode. The developed CM theory is also capable of computing the internal modal fields and modal radiation patterns.

In addition, a systematical feeding design methodology for DRAs is introduced through the investigation into the modal electric field and modal magnetic field of each DRA. The modal fields obtained from the CM analysis explicitly show how to excite the desired modes through carefully designing the feeding structures. The following are several important feeding design guidelines summarized through the examples we have introduced:

- If the desired mode of a DRA has very strong internal modal electric fields at a specific location in an elevation planes, it is possible to introduce the vertical coaxial probe at this location to excite the mode.
- The probe should extend through the ground plane until at the height at which the strong modal electric field vanishes in the elevation plane.
- If the desired mode of a DRA has very strong internal modal magnetic fields in the equatorial plane, it is possible to introduce a slot in the ground plane and feed the slot using microstrip line.
- The slot should be located at the location having strong modal magnetic fields, and the slot length has to be parallel with the direction of the magnetic field.

The above feeding design methodology is a physics-based approach. It is developed based on the natural electromagnetic properties of the dielectric resonators. Excitation of the dominant mode strictly at its resonant frequency ensures the good radiation efficiency and good input impedance matching. The excited radiation pattern approximates well to the modal radiation pattern. Moreover, it does not require the time-consuming "cut-and-try" process as in traditional design methods. It helps to reduce the design period and makes the DRA design experience much easier than ever.

Just like the contributions of the original CM theory [38] in many PEC antenna designs [51–56], the proposed CM theory for dielectric resonators pave a convenient and efficient way in the analysis, understanding, and design of the DRAs.

REFERENCES

[1] D. Richtinger, "Dielectric resonators," *J. Appl. Phys.*, vol. 10, pp. 391–398, Jun. 1939.

[2] A. Okaya and L. F. Barash, "The dielectric microwave resonator," *Proc. IRE*, vol. 50, pp. 2081–2092, Oct. 1962.

[3] S. A. Long, M. W. Mcallister, and L. C. Shen, "The resonant cylindrical dielectric cavity antenna," *IEEE Trans. Antennas Propag.*, vol. 31, no. 3, pp. 406–412, Mar. 1983.

[4] M. W. Mcallister and S. A. Long, "Rectangular dielectric-resonator antenna," *IEEE Electron. Lett.*, vol. 19, pp. 218–219, Mar. 1983.

[5] M. W. Mcallister and S. A. Long, "Resonant hemispherical dielectric antenna," *IEEE Electron. Lett.*, vol. 20, pp. 657–659, Aug. 1984.

[6] W. M. A. Wahab, D. Busuioc, and S.-N. Safieddin, "Low cost planar waveguide technology-based dielectric resonator antenna (DRA) for millimeter-wave applications: anal-

ysis, design, and fabrication," *IEEE Trans. Antennas Propag.*, vol. 58, no. 8, pp. 2499–2507, Aug. 2010.

[7] Y. M. Pan, K. W. Leung, and K.-M. Luk, "Design of the millimeter-wave rectangular dielectric resonator antenna using a higher-order mode," *IEEE Trans. Antennas Propag.*, vol. 59, no. 8, pp. 2780–2788, Aug. 2011.

[8] L. Ohlsson, T. Bryllert, C. Gustafson, et al., "Slot-coupled millimeter-wave dielectric resonator antenna for high-efficiency monolithic integration," *IEEE Trans. Antennas Propag.*, vol. 61, no. 4, pp. 1599–1607, Apr. 2013.

[9] M. J. Al-Hasan, T. A. Denidni and A. R. Sebak, "Millimeter-wave EBG-based aperture-coupled dielectric resonator antenna," *IEEE Trans. Antennas Propag.*, vol. 61, no. 8, pp. 4354–4357, Aug. 2013.

[10] K.-M. Luk and K.-W. Leung, *Dielectric Resonator Antennas*. Hertfordshire: Research Studies Press Ltd., p. 3, 2003.

[11] A. Petosa, *Dielectric Resonator Antenna Handbook*. Norwood, MA: Artech House, pp. 3, 2007.

[12] Bit-Babik, G., C. Di Nallo, and A. Faraone, "Multimode dielectric resonator antenna of very high permittivity," in IEEE Antennas & Propagation Symposium Digest AP-S 2004, Monterey, CA, vol. 2, pp. 1383–1386, Jun 2004.

[13] K. W. Leung, K. M. Luk, and K. Y. A. Lai, "Input impedance of a hemispherical dielectric resonator antenna," *IEE Electron. Lett.*, vol. 27, no. 24, pp. 2259–2260, Nov. 1991.

[14] K. W. Leung, K. M. Luk, K. Y. A. Lai, and D. Lin, "Theory and experiment of a coaxial probe fed hemispherical dielectric resonator antenna," *IEEE Trans. Antennas Propag.*, vol. 41, no. 10, pp. 1390–1398, Oct. 1993.

[15] G. P. Junker, A. A. Kishk, and A. W. Glisson, "Input impedance of dielectric resonator antennas excited by a coaxial probe," *IEEE Trans. Antennas Propag*, vol. 42, no. 7, pp. 960–966, Jul. 1994.

[16] A. Ittipiboon, R. K. Mongia, Y. M. M. Antar, M. Cuhaci, and P. Bhartia, "Aperture fed rectangular and triangular dielectric resonators for use as magnetic dipole antennas," *IEE Electron. Lett.*, vol. 29, no. 23, pp. 2001–2002, Nov. 1993.

[17] G. P. Junker, A. A. Kishk, and A. W. Glisson, "Input impedance of an aperture coupled dielectric resonator antenna," in IEEE Antennas & Propagation Symposium Digest, AP-S 1994, Seattle, WA, vol. 2, pp. 748–751, Jul. 1994.

[18] R. K. Mongia, A. Ittipiboon, Y. M. M. Antar, P. Bhartia, and M. Cuhaci, "A half-split cylindrical dielectric resonator antenna using slot-coupling," *IEEE Microw. Guid. Wave Lett.*, vol. 3, no. 2, pp. 38–39, Feb. 1993.

[19] A. Petosa, R. K. Mongia, A. Ittipiboona, and J. S. Wight, "Design of microstrip-fed series array of dielectric resonator antennas," *Electron. Lett.*, vol. 31, no. 16, pp. 1306–1307, Aug. 1995.

[20] A. Petosa, R. K. Mongia, A. Ittipiboona, and J. S. Wight, "Investigation of various feed structures for linear arrays of dielectric resonator antennas," in Antennas and Propagation Society International Symposium, 1995. AP-S. Digest, Newport Beach, CA, vol. 4, pp. 1982–1985, Jun. 1995.

[21] M. Jaworski and M. W. Pospieszalski, "An accurate solution of the cylindrical dielectric resonator problem," *IEEE Trans. Microw. Theory Tech.*, vol. MTT-27, no. 7, pp. 639–643, Jul. 1979.

[22] K. W. Leung and K. M. Luk, "Moment method solution of aperture-coupled hemispherical dielectric resonator antenna using exact modal Green's function," *IEE Proc.-Microw. Antennas Propag.*, vol. 141, no. 5, pp. 377–381, Oct. 1994.

[23] A. Petosa, "Chapter 2: Simple Shaped Dielectric Resonator Antennas," in Dielectric Resonator Antenna Handbook, Norwood, MA: Artech House, 2007.

[24] A. W. Glisson, D. Kajfez, and J. James, "Evaluation of modes in dielectric resonators using a surface integral equation formulation," *IEEE Trans. Microw. Theory Tech.*, vol. 31, no. 12, pp. 1023–1029, Dec. 1983.

[25] D. Kajfez, A. W. Glisson, and J. James, "Computed modal field distributions for isolated dielectric resonators," *IEEE Trans. Microw. Theory Tech.*, vol. 32, no. 12, pp. 1609–1616, Dec. 1984.

[26] Y. Liu, S. Safavi-Naeini, S. K. Chaudhuri, and R. Sabry, "On the determination of resonant modes of dielectric objects using surface Integral equations," *IEEE Trans. Antennas Propag.*, vol. 52, no. 4, pp. 1062–1069, Apr. 2004.

[27] R. Harrington, J. Mautz, and Y. Chang, "Characteristic modes for dielectric and magnetic bodies," *IEEE Trans. Antennas Propag.*, vol. 20, no. 2, pp. 194–198, Mar. 1972.

[28] Y. Chang and R. Harrington, "A surface formulation for characteristic modes of material bodies," *IEEE Trans. Antennas Propag.*, vol. 25, no. 6, pp. 789–795, Nov. 1977.

[29] A. J. Poggio and E. K. Miller, "Chapter 4: Integral equation solution of three dimensional scattering problems," in *Computer Techniques for Electromagnetics* (R. Mittra, ed.), Oxford, NY: Pergamon Press, 1973.

[30] H. Alroughani, J. L. T. Ethier, and D. A. McNamara "Observations on computational outcomes for the characteristic modes of dielectric objects," in 2014 IEEE Antenna and Propagation Symposiums, Memphis, TN, pp. 844–845, Jul. 6–11, 2014.

[31] Y. Chen and C.-F. Wang, "Surface integral equation based characteristic mode formulation for dielectric resonators," in IEEE Antennas Propagation Society AP-S International Symposium, Memphis, pp. 846–847, Jul. 2014.

[32] Y. Chen and C.-F. Wang, "Dual-band directional/omni-directional liquid dielectric resonator antenna designs using characteristic modes," in IEEE Antennas Propagation Society AP-S International Symposium, Memphis, pp. 848–849, Jul. 2014.

[33] R. F. Harrington, "Boundary integral formulations for homogeneous material bodies," *J. Electromagn. Waves Appl.*, vol. 3, no. 1, pp. 1–15, 1989.

[34] P. A. Raviart and J. M. Thomas, "A mixed finite element method for 2nd order elliptic problems," in Mathematical Aspects of Finite Element Methods, Proceedings Conference, Consiglio Naz. delle Ricerche (C.N.R.), Rome, Italy, pp. 292–315, Dec. 1975.

[35] S. M. Rao, D. Wilton, and A. W. Glisson, "Electromagnetic scattering by surfaces of arbitrary shape," *IEEE Trans. Antennas Propag.*, vol. 30, no. 3, pp. 409–418, May 1982.

[36] R. J. Garbacz, "Modal expansions for resonance scattering phenomena," *Proc. IEEE*, vol. 53, no. 8, pp. 856–864, Aug. 1965.

[37] K. Demarest and R. Garbacz, "Anomalous behavior of near fields calculated by the method of moments," *IEEE Trans. Antennas Propag.*, vol. 27, no. 5, pp. 609–614, Sep. 1979.

[38] R. Harrington and J. Mautz, "Theory of characteristic modes for conducting bodies," *IEEE Trans. Antennas Propag.*, vol. 19, no. 5, pp. 622–628, Sep. 1971.

[39] K. K. Chow, "On the solution and field patterns of cylindrical dielectric resonators," *IEEE Trans. Microw. Theory Tech.*, vol. 14, no. 9, pp. 439, Sep. 1966.

[40] R. K. Mongia and P. Bhartia, "Dielectric resonator antennas—a review and general design relations for resonant frequency and bandwidth," *Int. J. Microw. Millimeter-Wave Comput. Aided Eng.*, vol. 4, no. 3, pp. 230–247, Jul. 1994.

[41] R. K. Mongia, "Theoretical and experimental resonant frequencies of rectangular dielectric resonators," *IEE Proc. H Microwav. Antennas Propag.*, vol. 139, no. 1, pp. 98–104, Feb. 1992.

[42] K. W. Leung, K. Y. A. Lai, K. M. Luk, and D. Lin, "Input impedance of aperture coupled hemispherical dielectric resonator antenna," *Electron. Lett.*, vol. 29, no. 13, pp. 1165–1167, Jun. 1993.

[43] E. A. C. Marcatili, "Dielectric rectangular waveguide and directional coupler for integrated optics," *Bell Syst. Tech. J.*, vol. 48, no. 7, pp. 2071–2102, Sep. 1969.

[44] R. K. Mongia, and A. ittipiboon, "Theoretical and experimental investigations on rectangular dielectric resonator antennas," *IEEE Trans. Antennas Propag.*, vol. 45, no. 9, pp. 1348–1356, Sep. 1997.

[45] H. Y. Lo, K. W. Leung, K. M. Luk, and E. K. N. Yung, "Low profile equilateral-triangular dielectric resonator antenna of very high permittivity," *Electron. Lett.*, vol. 29, no. 23, pp. 2001–2002, Nov. 1993.

[46] A. A. Kishk, "A triangular dielectric resonator antenna excited by a coaxial probe," *Microw. Opt. Technol. Lett.*, vol. 30, no. 5, pp. 340–341, Sep. 2001.

[47] Y. Akaiwa, "Operation modes of a waveguide Y circulator," *IEEE Trans. Microw. Theory Tech.*, vol. 22, no. 11, pp. 954–960, Nov. 1976.

[48] X. S. Fang, K. W. Leung, and R. S. Chen, "On the wideband notched rectangular dielectric resonator antenna," in 2011 International Workshop on Antenna Technology (iWAT), Hong Kong, pp. 267–270, Mar. 7–9 2011.

[49] M. Cabedo-Fabres, E. Antonio-Daviu, A. Valero-Nogueira, and M. F. Bataller, "The theory of characteristic modes revisited: a contribution to the design of antennas for modern applications," *IEEE Antennas Propag. Mag.*, vol. 49, no. 5, pp. 52–68, Oct. 2007.

[50] J. J. Adams and J. T. Bernhard, "A modal approach to tuning and bandwidth enhancement of an electrically small antenna," *IEEE Trans. Antennas Propag.*, vol. 59, no. 4, pp. 1085–1092, Apr. 2011.

[51] D. Liu, R. J. Garbacz, and D. M. Pozar, "Antenna synthesis and optimization using generalized characteristic modes," *IEEE Trans. Antennas Propag.*, vol. 38 no. 6, pp. 862–868, Jun. 1990.

[52] Y. Chen and C.-F. Wang, "Electrically small UAV antenna design using characteristic modes," *IEEE Trans. Antennas Propag.*, vol. 62, no. 2, pp. 535–545, Feb. 2014.

[53] Y. Chen and C.-F. Wang, "Synthesis of reactively controlled antenna arrays using characteristic modes and DE algorithm," *IEEE Antennas Wirel. Propag. Lett.*, vol. 11, pp. 385–388, 2012.

[54] Y. Chen and C.-F. Wang, "Characteristic-mode-based improvement of circularly polarized U-slot and E-shaped patch antennas," *IEEE Antennas Wirel. Propag. Lett.*, vol. 11, pp. 1474–1477, 2012.

[55] Y. Chen and C.-F. Wang, "Electrically loaded Yagi-Uda antenna optimizations using characteristic modes and differential evolution," *J. Electromagn. Waves Appl.*, vol. 26, no. 8–9, 2012.

[56] Y. Chen and C.-F. Wang, "Synthesis of platform integrated antennas for reconfigurable radiation patterns using the theory of characteristic modes," in 10th International Symposium Antennas, Propagation EM Theory (ISAPE), Xian , pp. 281–285, Oct. 2012.

5

CHARACTERISTIC MODE THEORY FOR *N*-PORT NETWORKS

5.1 BACKGROUNDS

The characteristic mode (CM) theories for perfect electric conductors, for objects in multilayered medium, and for dielectric objects are discussed in Chapters 2, 3, and 4, respectively. In these CM theories, the characteristic currents exist as the continuous surface currents over the entire equivalent surface of the objects. These characteristic currents can be excited effectively through developing and placing feeding structures properly. In turn, the characteristic currents themselves provide valuable information for the design of these feeding structures. These CM theories are more suitable to electromagnetic objects in electrical small and medium sizes.

Alternatively, characteristic modes can be also defined for electrically large multiport antenna systems. Such multiport antenna systems can be generally regarded as the *N*-port networks, and the corresponding CM theory is called CM theory for *N*-port networks [1]. In this circumstance, the characteristic modes are determined from the *N*-port network's impedance matrix. The number of characteristic modes is equal to the number of the ports. The characteristic fields hold the orthogonality property over the radiation sphere, just like in the CM theories in Chapters 2, 3, and 4. Because of this attractive property, the CM theory for *N*-port networks is extensively applied in the analysis and design of multiport antenna systems [2–4], antenna arrays [5–9], and *N*-port loaded scatterers [10]. The goals in these studies were to obtain designated radiation pattern/scattering pattern over desired frequencies

Characteristic Modes: Theory and Applications in Antenna Engineering, First Edition.
Yikai Chen and Chao-Fu Wang.
© 2015 John Wiley & Sons, Inc. Published 2015 by John Wiley & Sons, Inc.

or to achieve desired impedance properties at the ports. Both of the far-field radiation performance and the impedance property are essentially determined by the radiating currents over the antennas or scatterers. Therefore, techniques based on the characteristic currents on the ports were developed to shape the current distributions over the antennas or scatterers. In particular, the following design goals were achieved by shaping the current distributions using various CM-based techniques:

- Realized mode decoupling elements designs at multiple input multiple output antenna ports to decouple the mutual couplings [2, 3];
- Achieved broadband impedance bandwidth by loading the ports with non-Foster elements [4];
- Achieved various far-field radiation performances (e.g., power pattern, gain, and sidelobe level (SLL)) by loading the ports with lumped reactive elements [5, 7, 8, 10];
- Synthesized power patterns by controlling the port currents using the far-field orthogonality property [6];
- Achieved ultra-wideband antenna arrays by controlling the port excitations [9].

In the following sections, we will first look at the fundamental theory of the characteristic modes for *N*-port networks in Section 5.2. Important properties of the CM theory will be discussed, albeit they are similar to those in the conventional CM theory [11]. Continuing with the fundamental theory discussions, technical details on the CM theory–based antenna array designs, including the beam scanning circular array, Yagi-Uda antennas and tightly coupled ultra-wideband antenna array, are then discussed through Sections 5.3–5.5. Finally, Section 5.6 concludes this chapter.

5.2   CHARACTERISTIC MODE FORMULATIONS FOR *N*-PORT NETWORKS

The CM theory for *N*-port networks was first developed by Harrington and Mautz in 1973 [1]. The computation of the characteristic modes is similar to the computation of the conventional characteristic modes for perfect electric conducting (PEC) objects [11]. While the conventional CMs are derived from the method of moment (MoM) impedance matrix of the PEC objects, the CMs for *N*-port networks are computed from the *N*-port network's impedance matrix. The impedance matrix for an *N*-port network can be obtained in many different ways, including full-wave electromagnetic simulations and measurement using a vector network analyzer. In the following, we first give the definitions for the impedance matrix of an *N*-port network.

Considering an *N*-port network as shown in Figure 5.1, an impedance matrix is defined to relate the voltages and currents at each port. Here, the "port" is defined as the access terminal planes at which the voltages and currents are defined. The ports in Figure 5.1 are of many types, such as a voltage gap on a metallic plane or a transmission line connected to the radiator. Incident waves (V_n^+, I_n^+) and reflected waves $(V_n^-, -I_n^-)$ are calculated or measured at the terminal planes. As can be observed

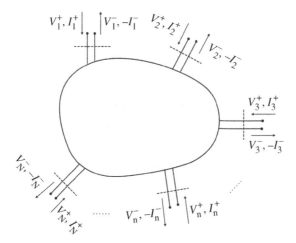

FIGURE 5.1   A general *N*-port network.

from Figure 5.1, the total voltage and current at the *n*th terminal plane are given by the following:

$$V_n = V_n^+ + V_n^-$$ (5.1)

$$I_n = I_n^+ - I_n^-$$ (5.2)

Therefore, the impedance matrix that relates the total port voltages and currents can be written as follows:

$$[V] = [Z][I]$$ (5.3)

where $[V] = [V_1, V_2, \ldots, V_N]^T$, $[I] = [I_1, I_2, \ldots, I_N]^T$, and

$$[Z] = \begin{bmatrix} Z_{11} & Z_{12} & \cdots & Z_{NN} \\ Z_{21} & Z_{22} & \cdots & Z_{2N} \\ \cdots & \cdots & \cdots & \cdots \\ Z_{N1} & \cdots & \cdots & Z_{NN} \end{bmatrix}$$ (5.4)

From Equation (5.3), we can find that

$$Z_{ij} = \frac{V_i}{I_j}\bigg|_{I_n = 0,\, n \neq j}$$ (5.5)

Equation (5.5) states that Z_{ij} is obtained by driving port j with the current I_j, keeping all the other ports to be open-circuit ($I_n = 0$, $n \neq j$), and calculating or measuring the open-circuit voltage V_i at the port i. Therefore, Z_{ii} is the input impedance looking into port i when all the other ports are open-circuit. Z_{ij} is the mutual impedance between ports i and j when all the other ports are open-circuit.

The impedance matrix $[Z]$ is generally a complex matrix for an arbitrary N-port network. If the system does not contain any active devices or non-reciprocal material, $[Z]$ is a symmetrical matrix. The characteristic modes for an N-port network are defined in a similar way as those for PEC bodies and can be solved from the following generalized eigenvalue equation:

$$[X]\bar{J}_n = \lambda_n [R]\bar{J}_n \tag{5.6}$$

where $[R]$ and $[X]$ are the real and imaginary parts of $[Z]$, that is,

$$[Z] = [R] + j[X] \tag{5.7}$$

$\bar{J}_n$ is the nth characteristic current at the ports, and λ_n is the corresponding eigenvalue of the nth characteristic mode. For antenna systems include only PEC structures and piecewise homogeneous materials, $[Z]$ is a symmetric matrix and $[R]$ is a real symmetric positive definite matrix. Most of the commonly used antennas satisfy these conditions. Therefore, all the eigenvalues λ_n are real and all the eigenvectors $\bar{J}_n$ are real vectors. The real eigenvector $\bar{J}_n$ indicates that the characteristic currents at all of the ports are equiphasal.

For convenience, the characteristic currents are normalized such that each of them delivers unit power:

$$\left(\bar{J}_n\right)^{\mathrm{T}} [R]\bar{J}_n = 1 \tag{5.8}$$

where $(\cdot)^{\mathrm{T}}$ denotes the transpose of the vectors. The orthogonality in the normalized characteristic currents are given by:

$$\left(\bar{J}_m\right)^{\mathrm{T}} [R]\bar{J}_n = \delta_{mn} \tag{5.9}$$

$$\left(\bar{J}_m\right)^{\mathrm{T}} [X]\bar{J}_n = \delta_{mn}\lambda_n \tag{5.10}$$

$$\left(\bar{J}_m\right)^{\mathrm{T}} [Z]\bar{J}_n = \delta_{mn}(1 + j\lambda_n) \tag{5.11}$$

where,

$$\delta_{mn} = \begin{cases} 1 & m = n \\ 0 & m \neq n \end{cases} \tag{5.12}$$

The characteristic currents form a set of basis to express the radiating currents on the N-port antenna system [1], that is, the modal solution of the characteristic modes for N-port network:

$$\bar{J} = -\sum_{n=1}^{N} \frac{\left(\bar{J}_n\right)^{\mathrm{T}} \bar{V}_{\mathrm{oc}}}{(1 + j\lambda_n)} \bar{J}_n \tag{5.13}$$

where $\bar{V}_{oc}$ is the open-circuit voltage vector that consists of the voltage for each port. Similar to the conventional CM theory for PEC objects, the modal solution in Equation (5.13) has two important components to determine the modal expansion coefficient. The denominator depends on the eigenvalue λ_n, which is determined by the *N*-port antenna system's configuration and materials. The numerator is determined by an inner product of the characteristic currents with the open-circuit voltage at each port. It shows the coupling strength between the modal currents and the external voltage source. As can be seen, the modal significance can be defined to show the intrinsic radiation capability of each mode. It is given by:

$$MS = \frac{1}{\left|1 + j\lambda_n\right|} \tag{5.14}$$

The modal significance maps the eigenvalue range $[-\infty, +\infty]$ into a much smaller range $[0, 1]$. It is more convenient to investigate the resonant behavior of high order modes over a wide frequency range.

Considering the contributions from the modal significance and the couplings between the characteristic currents and the voltage sources, the modal solution in Equation (5.13) can be reduced to:

$$\bar{J} = \sum_{n=1}^{N} a_n \bar{J}_n \tag{5.15}$$

where a_n is a complex modal expansion coefficient for the *n*th mode.

The electric fields and magnetic fields produced by the characteristic currents are called the characteristic fields. They form a Hilbert space of the field radiated by any practical currents on the ports. Similar to the conventional CM theory for PEC objects, the orthogonality relationships for the characteristic fields of *N*-port network can be also derived from the complex Poynting's theorem, and they are given by:

$$\frac{1}{\eta_0} \oiint_S \bar{E}_m \cdot \bar{E}_n^* dS = \delta_{nm} \tag{5.16}$$

$$\eta_0 \oiint_S \bar{H}_m \cdot \bar{H}_n^* dS = \delta_{nm} \tag{5.17}$$

where η_0 is the wave impedance in free space. As we shall see in the following sections, the orthogonality property of the characteristic fields is helpful in the analysis and synthesis of multiport radiation system or scattering systems. While keeping accuracy in the full-wave electromagnetic simulations, it greatly improves the efficiency in the synthesis of radiation patterns or scattering patterns.

5.3   REACTIVELY CONTROLLED ANTENNA ARRAY DESIGNS USING CHARACTERISTIC MODES

Reactively controlled antenna array (RCAA) is attractive in wireless communication systems due to its many distinct advantages including simple feed port configuration, high reliability, small power consumption, and low cost [5, 12–16]. RCAA is an *N*-port radiating system consisting of one element connected to an RF port and a number of surrounding parasitic elements with reactive loadings that can be realized by reversely biased varactor diodes.

The RCAA has been designed using various analysis and optimization techniques. As reported in Refs. [5, 12], optimum seeking univariate search method (USM) with the CM theory can control the direction of the main beam of a seven-element circular array by using reactance loadings. The limitation in this approach comes from the univariate search method. It cannot perform well for arrays with larger number of elements due to its slow convergence. In Ref. [14], a 25-element circular array with capacitive loadings has been optimized manually by repeating the simulations to get the near optimal radiation performance. The Monte Carlo method was also employed to get the optimal reactance loadings for reactively steered ring antenna arrays [15]. Later, the genetic algorithms (GA) [12] and direct search method [16] have been used to determine the capacitors loadings for circular arrays.

As described in Refs. [13–16], a common serious issue is that they demand full-wave simulations to evaluate the performance of each possible candidate, such that the optimizer can determine the search direction in the next seeking step. It often makes the optimization process very computational costly.

This section discusses the reactively loaded beam scanning circular array designs using the characteristic mode theory for *N*-port networks [7]. A simple yet efficient approach for the synthesis of reactively controlled antenna arrays is proposed by combing the characteristic mode theory with the differential evolution (DE) algorithm [17]. In this approach, characteristic mode analysis for the *N*-element circular antenna array is conducted first to obtain various radiating modes. Based on these characteristic modes, the DE algorithm is then employed to seek the optimal mode excitation coefficients according to the specific requirements for the radiation pattern, loading schemes, and/or other specifications. Several circular antenna arrays are successfully synthesized according to main beam direction and maximum SLL. These antenna array designs prove that the characteristic modes for the *N*-port network give important decisions to the optimal loading designs for antenna arrays. These examples also demonstrate the effectiveness of the proposed synthesis method based on characteristic modes and DE algorithm.

5.3.1   Problem Formulation

Without loss of generality, we consider the reactively controlled antenna array in Figure 5.2. There are six passive half-wavelength dipole antenna elements surrounding an active element. The active dipole antenna is located at the center of the circular array. It is actively fed by an RF cable. The surrounding antenna elements are indirectly fed through the electromagnetic coupling from the center dipole antenna.

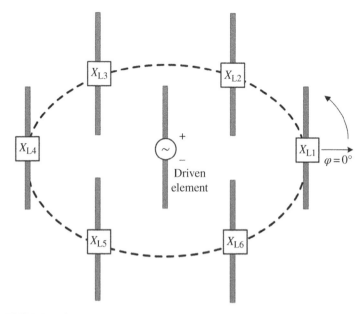

FIGURE 5.2   A seven-element reactively controlled circular antenna array.

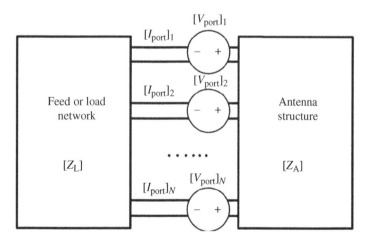

FIGURE 5.3   The Thevenin equivalence of an N-port reactively controlled antenna array.

The Thevenin equivalence of a general N-port reactively controlled antenna array is shown in Figure 5.3. As can be seen, the terminal impedance characteristics of the antenna array are represented by an impedance matrix $[Z_A]$, and the terminal impedance characteristics of the loading/excitation network is described by a diagonal impedance matrix $[Z_L]$.

The $[Z_A]$ can be computed from various electromagnetic numerical techniques such as the MoM [18] and the finite element method [19, 20]. In practice, it can be also measured using a vector network analyzer. In this chapter, the impedance matrix $[Z_A]$ is computed from the MoM. Specifically, by applying a unit voltage source at each port when keeping all the other ports open-circuit, the current at each port can be solved using the MoM. It results in a matrix $\left[I_{port}\right]_{N\times N}$ by applying the unit voltage source to each element. Therefore, the voltage matrix can be expressed in terms of $[Z_A]$ and $\left[I_{port}\right]_{N\times N}$:

$$\left[Z_A\right]_{N\times N}\cdot\left[I_{port}\right]_{N\times N}=\left[V_{port}\right]_{N\times N} \tag{5.18}$$

Because unit voltage source is applied to each port, the voltage matrix is an identity matrix. Therefore, we can get the impedance matrix $[Z_A]$ by inversing the matrix $\left[I_{port}\right]_{N\times N}$:

$$\left[Z_A\right]_{N\times N}=\left[I_{port}\right]_{N\times N}^{-1} \tag{5.19}$$

Evidently, if a loading/excitation network is connected to the multiport antenna system, the port current should be computed from a new matrix equation [1],

$$\left[Z_A+Z_L\right]_{N\times N}\left[I_{port}\right]_{N\times N}=\left[V_{port}\right]_{N\times N} \tag{5.20}$$

The goal of the reactively controlled antenna array design is to find a set of reactance loadings such that the beam can steer to a designated direction and the maximum SLL can be suppressed below a given level.

In conventional methods, the optimization is directly applied to seek the optimal loadings $[Z_L]$. The currents over the antenna system (not only the port currents) have to be solved from the MoM matrix equation repeatedly such that the radiation performance can be evaluated in the optimization procedure. However, solving the MoM matrix equation is quite time-consuming. The conventional method thus may fail if antenna arrays with large number of elements are considered.

5.3.2   Design and Optimization Procedure

In order to avoid the solution of the MoM matrix equation in each step of the cost function evaluation, a new synthesis method based on the CM theory and DE algorithm is developed. According to the CM theory, any currents $\bar{J}$ at the ports of a multiport antenna system can be expanded using the characteristic currents $\bar{J}_n$:

$$\bar{J}=\sum_{n=1}^{N}a_n\bar{J}_n \tag{5.21}$$

where a_n are the coefficients to be determined for designated radiation performance of the RCAA. Similarly, the radiation field $\bar{E}$ radiated by $\bar{J}$ can be also written as the superposition of the N characteristic field $\bar{E}_n$ that is produced by the characteristic currents $\bar{J}_n$:

$$\bar{E}=\sum_{n=1}^{N}a_n\bar{E}_n \tag{5.22}$$

As can be seen, in the framework of the CM theory, designing an RCAA to meet some particular specifications can be considered as the solution of a nonlinear optimization problem. The goals of the array synthesis problem are to steer the maximum gain direction to a desired direction φ_0, maximize the front-to-back (F/B) ratio of the radiation pattern, and suppress the maximum SLL by tuning the modal expansion coefficients a_n. After getting the optimal set of modal expansion coefficients, the reactive loadings (diagonal elements in the matrix $[Z_L]$) can be determined directly from a_n.

In general, any optimization algorithm can be applied to seek the optimal set of modal expansion coefficients a_n. As compared to the optimum seeking USM [5], population-based evolutionary algorithms, such as GA [21], particle swarm optimization (PSO) algorithm [22], are more powerful in solving nonlinear optimization problems. Recently, DE algorithm has been widely applied to solve many electromagnetic problems [17, 23]. It has been proven that the DE algorithm has excellent search ability and faster convergence rate over most of the optimization algorithms such as the GA and the PSO methods. It is especially suitable to real-valued optimization problems. The main procedure of the algorithm is described as follows.

Figure 5.4 shows the flowchart of the DE algorithm. Upon the solution space, cost function (or objective function) and other fundamental parameters are defined, and the DE algorithm starts with the initialization of the evolutionary population. We apply the uniform probability distribution in the population initialization. Meanwhile, we assume all the random decisions in the DE algorithm follow the uniform probability distribution. For an optimization problem with N_{PAR} parameters, the population can be written as a $N_{POP} \times N_{PAR}$ matrix:

$$V = \begin{bmatrix} v_{1,1} & \cdots & v_{1,N_{PAR}} \\ \cdots & \cdots & \cdots \\ v_{N_{POP},1} & \cdots & v_{N_{POP},N_{PAR}} \end{bmatrix} \tag{5.23}$$

where N_{POP} is the number of individuals in the population. It remains unchanged during the optimization. $v_{i,j}$ is the jth gene of the ith individual. Each row of the matrix represents a candidate solution to the optimization problem.

The crucial idea behind the DE algorithm is the scheme for generating the trial parameter vectors. The DE algorithm generates new parameter vectors by adding a weighted differential vector between two population members to a third member. In this work, the third member is chosen to be the best individual in the current population to enhance the greediness of the algorithm. Thus, the mutant vector can be written as:

$$v^{M,i} = v^{(n),opt} + \beta \left(v^{(n),p_1} - v^{(n),p_2} \right), \quad i \neq p_1 \neq p_2 \neq opt \tag{5.24}$$

where the integer p_2 and p_2 are chosen randomly from the interval $[1, N_{POP}]$ and are different from the running index i. The superscript opt denotes the current best

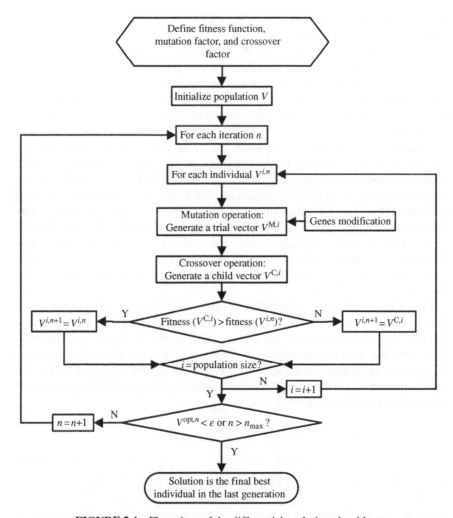

FIGURE 5.4   Flow chart of the differential evolution algorithm.

vector in the parent population at the *n*th generation. The mutation factor β is a real and constant factor introduced to control the amplification of the differential variation.

The genes can be modified in the following way when they exceed the predefined search range:

$$
\left(v^{M,i}\right)_j = \begin{cases} \dfrac{\left(v^{M,i}\right)_j + B^i}{2}, & \left(v^{M,i}\right)_j < A^i \\[3mm] \dfrac{\left(v^{M,i}\right)_j + A^i}{2} & \left(v^{M,i}\right)_j > B^i \end{cases} \tag{5.25}
$$

where j is the gene index and $[A^i, B^i]$ is the search range of the ith optimization parameter.

In order to increase the diversity of the parameter vector in the population, the child vector v^C is selected according to the following rule:

$$\left(v^{C,i}\right)_j = \begin{cases} \left(v^{M,i}\right)_j & \gamma < p_{\text{cross}} \\ \left(v^{(n),i}\right)_j & \text{otherwise} \end{cases} \tag{5.26}$$

where γ is a real random number in the range [0, 1], and p_{cross} is the crossover factor. Through this operation called as crossover, a certain sequence of the vector elements of $v^{C,i}$ are identical to the elements of $v^{M,i}$, while the other elements of $v^{C,i}$ directly inherit from $v^{(n),i}$.

In the final step, the selection operation takes place. If the child vector has a cost function value smaller than its parent vector, the newly generated child vector replaces its parent vector. The overall procedure repeats until the maximum number of generations are used or the iteration accuracy is satisfied.

For any global optimization algorithm, inappropriate selection of the control parameters may significantly deteriorate the robustness and greediness. The DE algorithm have three important intrinsic control parameters, including the mutation factor β, the crossover probability p_{cross}, and the population size N_{POP}. They play important roles in the DE algorithm, and only properly selected values of them can ensure a fast and robust convergence for DE. Empirically, the typical values of them are selected as follows [17]: $0.5 \le \beta \le 0.7$, $0.7 \le p_{\text{cross}} \le 0.9$, and N_{POP} is approximately four or five times of the number of optimization parameters, that is, $N_{\text{POP}} = (4 \sim 5)N_{\text{PAR}}$.

To apply the DE algorithm to find the optimal modal expansion coefficients α_n, such that the main beam direction, the F/B ratio of the radiation pattern, and the SLL meet the specified value, construction of a good cost function is of great importance. The cost function ensures the DE optimizer will converge to the desired antenna radiation performance. The following cost function is defined for the reactively controlled beam steering antenna array synthesis problem:

$$\begin{aligned} F(v) = & w_1 \left| \phi_0 - \phi_0^{(k)} \right| + w_2 E\left(\phi_{\text{backward}}^{(k)} \right) \\ & + w_3 \left| \text{SLL}^{(k)} - \text{SLL}_{\text{desired}} \right| U\left(\text{SLL}^{(k)} - \text{SLL}_{\text{desired}} \right) \\ & + w_4 \text{real}\left(Z_{\text{Li}}^{(k)} \right) \end{aligned} \tag{5.27}$$

where the superscript k denotes the evolution generation index and w_i, $i = 1, 2, 3, 4$ are the weighting factors to balance the contributions of each term to the cost function. φ_0 and $\varphi_0^{(k)}$ are the desired and calculated main beam direction at the kth generation, respectively. $\phi_{\text{backward}}^{(k)}$ is the intensity of the electricity field at the backward direction that is defined with respect to the main beam direction φ_0. $\text{SLL}_{\text{desired}}$ and $\text{SLL}^{(k)}$ are the desired and the maximum SLL at the kth generation, respectively. $U(\cdot)$ is a step function, that is, $U = 1$ for $\text{SLL}^{(k)} > \text{SLL}_{\text{desired}}$ and $U = 0$ for $\text{SLL}^{(k)} \le \text{SLL}_{\text{desired}}$.

The term $Z_{Li}^{(k)}$ is included in the cost function to enforce the real part of the impedance loading to be zero to allow purely reactive loadings.

For the reactively controlled antenna array in Figure 5.2, only the central element is directly excited, and the surrounding elements are loaded with reactance. The voltages implemented at the port of the all the elements can be expressed as:

$$\left[V_{\text{port}}\right] = \left[0, \ 0, \ 0, \ 0, \ 0, \ 0, \ v_s - Z_0 I_7\right]^{\text{T}} \tag{5.28}$$

where $Z_0 = 50 \ \Omega$ is the characteristic impedance of the RF cable that directly feeds the central active element. v_s and I_7 are the voltage source and the current flows across the port at the central element, respectively. Therefore, the last element in the vector $[V_{\text{port}}]$ denotes the real voltage across the port of the central element. The impedance matrix $[Z_L]$ for the loading/excitation network is a diagonal matrix and is given by:

$$[Z_L] = \begin{bmatrix} jX_{L1} & 0 & 0 & 0 & 0 & 0 & 0 \\ 0 & jX_{L2} & 0 & 0 & 0 & 0 & 0 \\ 0 & 0 & jX_{L3} & 0 & 0 & 0 & 0 \\ 0 & 0 & 0 & jX_{L4} & 0 & 0 & 0 \\ 0 & 0 & 0 & 0 & jX_{L5} & 0 & 0 \\ 0 & 0 & 0 & 0 & 0 & jX_{L6} & 0 \\ 0 & 0 & 0 & 0 & 0 & 0 & 0 \end{bmatrix} \tag{5.29}$$

where jX_{Li} $(i = 1,2,\ldots,6)$ are the reactance loadings on the parasitic elements. The last diagonal element in $[Z_L]$ is equal to zero. It indicates that there is no reactive loading applied on the central active element.

When the optimization algorithm yields a set of optimal candidate solution α_n $(n = 1,2,\ldots,N)$, the port currents $[I_{\text{port}}]$ can be computed directly using the modal solution in Equation (5.21). Because the impedance matrix $[Z_A]$ keeps unchanged and the voltage excitation $[V_{\text{port}}]$ in Equation (5.28) is known before optimization, the loading impedance on the parasitic elements only depends on the port currents $[I_{\text{port}}]$, and it can be directly computed from the following formulation:

$$Z_{Li} = \frac{\left(\left[V_{\text{port}}\right] - \left[Z_A\right]\left[I_{\text{port}}\right]\right)_i}{\left[I_{\text{port}}\right]_i}, \ i = 1,2,\ldots,6 \tag{5.30}$$

By substituting the load impedance Z_{Li} into Equation (5.27), the cost function for any design candidates $\alpha_n, n = 1,2,\ldots,6$ can be computed efficiently without solving the MoM matrix equation. However, the mutual coupling among the elements are taken into account in the optimization by using the characteristic currents and characteristic fields. This is also what makes the CM-based approach be different from those reported in Refs [13–16].

5.3.3 Design Examples

Three RCAAs are synthesized to verify the effectiveness of the CM-based approach. The design specifications include the main beam direction, F/B ratio, and maximum SLL. A well-verified in-house MoM code is applied to calculate the N-port network impedance matrix and embedded element pattern for the RCAAs.

As the first example, a seven-element circular array with one-quarter wavelength spacing is considered for the synthesis of steered radiation patterns with maximum F/B ratio. Half-wavelength dipoles of a diameter of $\lambda/200$ are used as the antenna elements. The six surrounding elements are intended to load with reactance X_{Li} across their input terminals. The center active element is excited by a unit voltage source $v_s = 1$ with an internal impedance of $Z_0 = 50\ \Omega$.

Because we only need to constrain the main beam direction, backward radiations, and the value of the loaded impedances in this design, the weighting factor w_3 in the cost function of Equation (5.27) is set to zero. Figure 5.5 shows the steered radiation patterns achieved by tuning the reactance loadings in the seven-element circular array. As can be seen, the beams are successfully steered to the directions of $\phi_0 = 30°, 45°$, and $90°$, respectively. Meanwhile, the backward radiations are suppressed to a very low level in all of the three beams. Table 5.1 gives the reactance values loaded on the surrounding elements. As can be seen, only purely reactance loadings are necessary to steer the beam in different directions. It indicates that the last term in Equation (5.27) indeed enforces the real part of the impedance Z_{Li} to be zero.

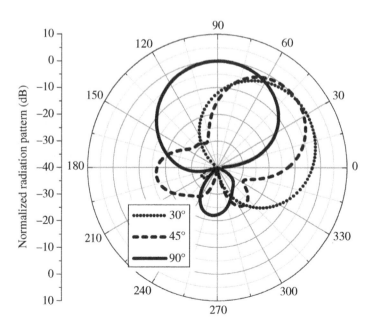

FIGURE 5.5 Normalized radiation patterns of the seven-element reactively loaded circular array. From Ref. [7]. © 2012 by IEEE. Reproduced by permission of IEEE.

TABLE 5.1   Reactance Loadings for the Steered Seven-element Circular Array

Beam angle	X_{L1}	X_{L2}	X_{L3}	X_{L4}	X_{L5}	X_{L6}
30°	−123.48	−122.95	91.57	28.16	24.49	85.90
45°	−98.82	−66.18	−153.41	−7.07	80.55	43.80
90°	71.92	−76.94	−81.65	189.75	21.27	86.45

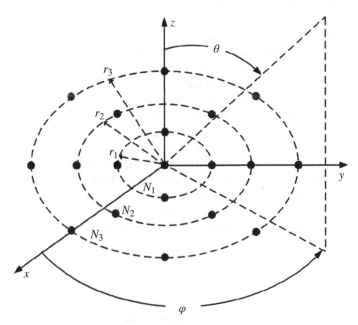

FIGURE 5.6   Configuration of the three-ring concentric circular antenna array ($N_1=4$, $N_2=6$, $N_3=8$).

As the second design case, radiation pattern with low SSL is expected from a three-ring concentric circular antenna array (CCAA). The same dipole element as in the first example is placed along three concentric circular rings. Figure 5.6 shows the configuration of the CCAA. The CCAA is assumed to be with M rings, the radius of the mth ($m = 1,2,...,M$) ring is r_m and N_m is the number of elements placed on the mth ring. The elements on each ring are assumed to be uniformly distributed with angles $\Phi_{mi} = 2\pi\left[(i-1)/N_m\right]$, $m = 1,2,...,M$; $i = 1,2,...,N_m$ [24].

Because we only need to constrain the main beam direction, the maximum SLL, and the loaded impedance values, the weighting factor w_2 is set to zero in the cost function of Equation (5.27). Two CCAA is designed with the CM-based approach. The radiuses of the two CCAAs are the same, $(r_1,r_2,r_3) = (0.55\lambda,0.61\lambda,0.75\lambda)$, and the number of antenna elements on the three rings for the two CCAA are $(N_1,N_2,N_3) = (4,6,8)$ and $(8,10,12)$, respectively. For both of the two CCAA designs, we assume that the main beam has to point to the negative x-axis direction, that is, $\varphi_0 = 180°$. Figure 5.7 shows the normalized radiation patterns obtained using the

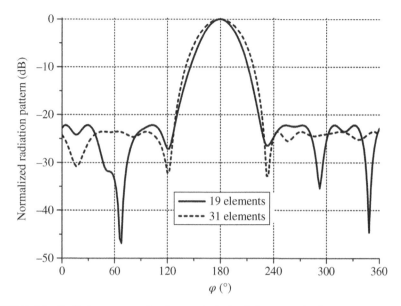

FIGURE 5.7 Radiation patterns for the reactively loaded concentric circular antenna arrays with 19 and 31 elements. From Ref. [7]. © 2012 by IEEE. Reproduced by permission of IEEE.

TABLE 5.2   Reactance Loadings for the Two Three-ring CCAAs

	X_{Li} for the 19-element array			X_{Li} for the 31-element array		
i	First ring	Second ring	Third ring	First ring	Second ring	Third ring
1	−115.48	54.53	41.93	−85.34	559.90	−15.92
2	−13.06	−166.89	−3.19	−33.82	117.14	10.49
3	−499.98	499.99	−67.95	599.93	46.98	−64.65
4	−14.41	−186.12	−186.47	−599.94	143.26	46.76
5	—	500.00	496.14	−289.22	−599.86	105.65
6	—	−172.96	−187.14	−322.04	−93.98	−87.46
7	—	—	−68.24	524.66	452.62	−452.69
8	—	—	−0.49	−54.99	66.88	−70.39
9	—	—	—	—	86.92	106.82
10	—	—	—	—	227.05	228.74
11	—	—	—	—	—	−35.92
12	—	—	—	—	—	26.32

optimal loaded reactance values given in Table 5.2. As can be observed from Figure 5.7, the obtained maximum SSL for the 19 (=4+6+8+1) and 31 (=8+10+12+1) element cases are −22.17 and −23.54 dB, respectively. It is evident that the SLL of the CCAAs is suppressed to a very low level by using the reactive loadings. As compared to conventional low-sidelobe antenna array designs, the present approach avoids the costly and complicated feeding networks.

5.3.4 Efficiency of CM-Based Approach

This section investigates the computational efficiency of the CM-based approach from the following two aspects: the convergence rate of the cost function and the CPU time.

To illustrate the convergence rate of the cost function, the cost function values of the best and worst individuals in the evolution procedures of the DE algorithm for the two CCAA design cases are shown in Figure 5.8. It can be observed that the initial design candidates of both cases have large cost function values, but they quickly converge to their optimal values after 10,000 cost function evaluations.

The CPU time for the cost function evaluations is also investigated for the two CCAA designs. Table 5.3 compares the required CPU time for 100,000 candidate performance evaluations in the case that the modal solutions of the CM theory and the MoM are implemented in the DE optimizer. It is evident that more than three orders of speedup is achieved in the CM-based approach. This speedup is due to the replacement of conventional full-wave simulations with the modal solution in the CM theory.

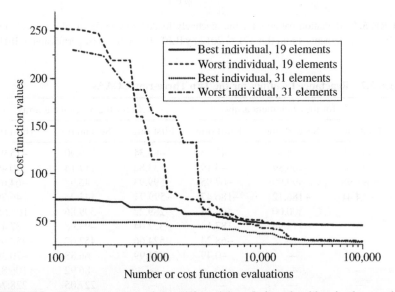

FIGURE 5.8 Convergence rate of the differential evolution algorithm in the reactively loaded concentric circular antenna array optimizations.

TABLE 5.3   CPU Time for Evaluation of 100,000 Objective Functions

Design cases	CPU time (s)	
	DE with theory of CM	DE with MoM
19 element CCAA	12.4	30,638.6
31 element CCAA	22.3	133,073.9

In summary, by combing the CM theory with the DE algorithm, reactively controlled circular antenna arrays are synthesized for achieving beam steering and low-sidelobe radiation patterns. Based on the characteristic mode analysis to the antenna array, the DE algorithm is applied to find the optimal modal expansion coefficients for each mode. The reactance loadings can be determined directly from the modal expansion coefficients. The array synthesis efficiency is greatly improved due to the modal solutions in the CM theory. The CM-based array synthesis approach is also a promising method for other antenna array designs, especially for those arrays consisting of large number of antenna elements.

5.4 YAGI-UDA ANTENNA DESIGNS USING CHARACTERISTIC MODES

Yagi-Uda antennas are widely used as high-gain antennas in the high frequency (HF), very high frequency (VHF), and ultra high frequency (UHF) bands [25]. It can be designed to meet radiation performances such as high F/B ratio and low SLLs. Moreover, it features with attractive merits such as lightweight, low cost, and easy to fabricate. Yagi-Uda antenna consists of one driven antenna element, one reflector antenna element, and one or more director antenna elements. The antenna characteristics are controlled by the currents distributed on the driven element and the induced currents on the parasitic elements through electromagnetic coupling [26].

In the early days, Ehrenspeck and Poehler [27] determined the optimal element lengths and spacings for maximum gain, according to their extensive laboratory experimental results. Later, Cheng made substantial efforts in the optimal Yagi-Uda antenna designs using several gradient-based optimization techniques [28–30]. They found that the gain of a Yagi-Uda antenna is a highly nonlinear function with respect to the physical dimensions. Based on their work on Yagi-Uda antenna designs, they presented the following conclusion: There are many minor maxima for directivity (or gain) in the multidimensional space. Therefore, Cheng's Yagi-Uda antenna designs were largely dependent on the choice of initial solution. Actually, it is not an easy task to provide a good initial design for distinct design objectives. As can be seen, it is difficult to develop closed form expressions for the determination of element lengths and spacings of Yagi-Uda antennas for particular design specifications.

In recent years, population-based evolutionary algorithms are widely used in Yagi-Uda antenna optimizations. As early as 1997, two independent research groups applied the GA for the design of Yagi-Uda antennas to achieve optimal antenna characteristics [31, 32]. As another attempt, Lohn solved the Yagi-Uda antenna optimization problems using a binary coded evolutionary algorithm [33]. Other evolutionary algorithms such as the computational intelligence (CI) [34], the comprehensive learning particle swarm optimization (CLPSO) [35], the simulated annealing algorithm (SA) [36], and the biogeography-based optimization (BBO) method [37] were also introduced for the optimal design of Yagi-Uda antennas.

These researches have well proved that population-based evolutionary algorithms are quite useful tools for optimization problems involving highly nonlinear and/or non-differentiable cost functions. Meanwhile, they are less prone to convergence to a local minimum. However, as compared to deterministic optimization techniques such as the gradient-based techniques [28–30], population-based evolutionary algorithms are often criticized for the following three reasons:

1. They are not mathematically rigorous. The optimization theory cannot ensure they can work for all of the design cases, although they are indeed efficient in a large number of designs.

2. They are computationally expensive. They require a large number of cost function evaluations. Each of the cost function evaluation includes a full-wave simulation running, which is usually very time-consuming. Both the large number of cost function evaluations and the full-wave analysis impose heavy computation burdens to the overall optimization procedure.

3. The optimization-based designs lack physical meanings. It is purely a numerical technique.

On the other hand, the performance of Yagi-Uda antennas heavily depends on sensitive physical dimensions such as element lengths and spacings. Requirements on the fabrication accuracy for the realization of optimal designs impose great difficulty in practice, especially for antennas operated at high frequencies. Alternatively, it may be possible to achieve optimal Yagi-Uda antenna designs by reactively loading the parasitic elements, instead of tuning the sensitive physical sizes.

In this chapter, the novel CM-based antenna array synthesis procedure is implemented in the design of reactively loaded Yagi-Uda antennas. By performing the CM analysis using the CM theory for *N*-port networks, Yagi-Uda antenna performances with different reactance loadings can be evaluated efficiently. The DE algorithm iteratively search the candidate solution in the decision space until the radiation performance satisfies the specifications on the gain, input impedance, maximum SLL, and so on. To confirm the capabilities and efficiency of the CM based approach, performances of the optimal Yagi-Uda antenna designs are compared with those optimized by the BBO method, the comprehensive learning PSO method, the CI method, and the GA.

5.4.1   CM-Based Design Method

Without loss of generality, we consider a reactively loaded six-element Yagi-Uda antenna that consists of one driven element, four director elements, and one reflector element. Figure 5.9 shows the configuration of the reactively loaded six-element Yagi-Uda antenna. In the current context, the element lengths L_i ($i = 1,2,...,6$) and element spacings S_i($i = 1,2,...,5$) are determined by empirical formulations or personal experiences. They are fixed in the optimization procedure and their initial values are easy to determine. Moreover, because the radiation performance is largely dependent on the reactive loadings, these fixed dimensions will not dominate the final antenna performances. As a general rule, the elements have to locate closely

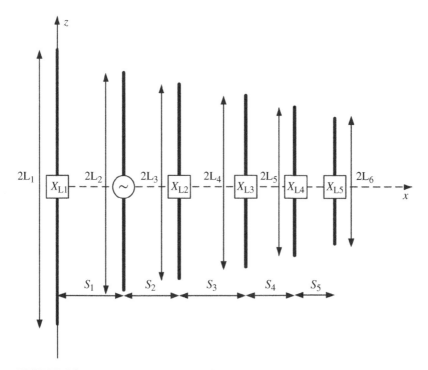

FIGURE 5.9 Configuration of the reactively loaded six-element Yagi-Uda antenna.

with each other such that the parasitic elements can be efficiently excited through the power coupling from the active elements. We suggest choosing the element spacings in the Yagi-Uda antennas to be around 0.25λ, where λ is the wavelength in free space [8].

In Figure 5.9, X_{Li} ($i = 1,2,...,5$) are the reactance loadings on the parasitic elements. They are the design variables in the DE optimizer. The design objectives include the antenna gains, the maximum SLLs, and the input impedances. They are highly nonlinear functions of these design variables. Although numerical techniques such as the method of moments are capable of computing the current distributions over reactively loaded antennas [18], computational burdens make it impossible to be directly implemented in optimization algorithms. Evidently, almost all of the Yagi-Uda antenna design methods have this issue. Therefore, the CM theory for N-port networks is introduced to achieve efficient analysis of reactively loaded Yagi-Uda antennas.

The fundamental theory of Yagi-Uda antennas states that the electromagnetic waves received or transmitted from the driven element introduce induced currents on the parasitic elements through electromagnetic coupling. The total radiation performance is determined by the driven element as well as the reactively loaded parasitic elements. The Thevenin equivalent circuit in Figure 5.3 is also applicable to the reactively loaded Yagi-Uda antenna. It is clear that if a loading/excitation network is connected to the antenna structure, the port current can be solved from the following equation:

$$\left[Z_A + Z_L\right]_{N \times N} \cdot \left[I_{port}\right]_{N \times 1} = \left[V_{port}\right]_{N \times 1} \tag{5.31}$$

As can be seen, the objective of the reactively loaded Yagi-Uda antenna design is to find a set of reactance loadings Z_L, such that the antenna performances meet the desired specifications.

Instead of directly optimizing the reactance loadings Z_L, the DE algorithm is applied to find the optimal modal expansion coefficients α_n for each characteristic mode. In order to determine the optimal values of α_n using the DE algorithm, the cost function is of great importance to ensure that the optimizer converges to the global optimal solution. The following cost function is thus constructed for the reactively loaded Yagi-Uda antenna design problems:

$$F(v) = -w_1 G^{(k)}(v) + w_2 \left|SLL^{(k)} - SLL_{desired}\right| U\left(SLL^{(k)} - SLL_{desired}\right)$$
$$+ w_3 \left|50 - Re\left(Z_{in}^{(k)}\right)\right| + w_4 \left|Im\left(Z_{in}^{(k)}\right)\right| \tag{5.32}$$

where k is the index for the evolution generations in the DE optimizer and w_i ($i = 1,2,3,4$) are the weighting factors for each term to balance the contributions of each term to the cost function. $G^{(k)}(v)$ is the optimal gain at the kth generation. $SLL_{desired}$ and $SLL^{(k)}$ are the desired and the optimized maximum SLL at the kth generation, respectively. $U(\cdot)$ is a step function to ensure the $SLL^{(k)}$ be lower than the prescribed $SLL_{desired}$. It is defined as $U = 1$ for $SLL^{(k)} > SLL_{desired}$ and $U = 0$ for $SLL^{(k)} " SLL_{desired}$. $Re\left(Z_{in}^{(k)}\right)$ and $Im\left(Z_{in}^{(k)}\right)$ are the real and imaginary parts of the input impedance of the Yagi-Uda antenna. The last two terms in Equation (5.32) enforce that the input impedance matches with the standard $50\,\Omega$ transmission line.

The modal solution of the CM theory given by Equation (5.15) indicates that the port current $\left[I_{port}\right]_{N \times 1}$ can be easily computed from the superposition of the weighted characteristic current at each port. For the reactively loaded Yagi-Uda antenna in Figure 5.9, only the driven element is excited, and the parasitic elements are loaded with reactance. Figure 5.10 shows the driven element that is excited by a voltage source with internal impedance. Therefore, the excitation voltages applied on the ports of the Yagi-Uda antenna can be written as:

$$\left[V_{port}\right] = \left[0, v_s - Z_0 I_2, 0, 0, 0, 0\right]^T \tag{5.33}$$

where $Z_0 = 50\ \Omega$ is the characteristic impedance of the RF cable that directly connected to the driven element, v_s is the voltage across the port of the driven element,

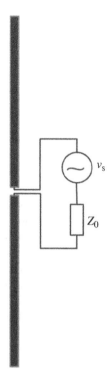

FIGURE 5.10 Voltage source for the excitation of the driven element.

and I_2 is the current flow across the port. The N-port impedance matrix $[Z_L]$ for the loading/excitation network is a diagonal matrix. It is given by:

$$[Z_L] = \begin{bmatrix} jX_{L1} & 0 & 0 & 0 & 0 & 0 \\ 0 & 0 & 0 & 0 & 0 & 0 \\ 0 & 0 & jX_{L2} & 0 & 0 & 0 \\ 0 & 0 & 0 & jX_{L3} & 0 & 0 \\ 0 & 0 & 0 & 0 & jX_{L4} & 0 \\ 0 & 0 & 0 & 0 & 0 & jX_{L5} \end{bmatrix} \qquad (5.34)$$

where jX_{Li} ($i = 1, 2, \ldots, 5$) are the loading reactances on the parasitic elements. Because there is no impedance loading applied in the driven element, the second diagonal element in $[Z_L]$ is set to zero.

Implementing simple mathematic manipulations in Equations (5.31), (5.33), and (5.34), we can calculate the loading reactances directly from the synthesized port currents $[I_{port}]$ and the loading impedance on the parasitic elements can be computed from:

$$Z_{Li} = \frac{\left([V_{port}] - [Z_A][I_{port}] \right)_i}{[I_{port}]_i}, \quad i = 1, 2, \ldots, 5 \qquad (5.35)$$

5.4.2 Design Examples

In this section, the CM-based approach is applied to synthesize four reactively loaded Yagi-Uda antennas. The synthesized antenna performances are compared with those obtained by directly optimizing the element lengths and spacings using the GA [31, 32] CI [34], the CLPSO [35], and the BBO [37]. The efficiency of the CM-based approach is also demonstrated by comparing with the conventional methods.

5.4.2.1 Six-Element Yagi-Uda Antenna Designs

A 6-element Yagi-Uda antenna with reactance loadings on the parasitic elements (including the reflector and directors) is designed for achieving maximum gain at the center frequency. We refer to it as the Yagi-Uda Antenna I in the following discussion. In this case, only the first term in the cost function of Equation (5.32) is used to maximize the antenna gain. Therefore, the weighting factors are set as follows: $w_1 = 1$, $w_i = 0$ ($i = 2,3,4$). The element lengths and spacings are determined from the general knowledge of Yagi-Uda antennas. The element lengths are chose to be around half-wavelength, and the element spacings are set to be around one-quarter wavelength. The element lengths and spacings for Yagi-Uda Antenna I are shown in Table 5.4. The element radiuses for all of the elements are 0.003369λ. The optimized reactances loaded on the parasitic elements of the Yagi-Uda Antenna I are also shown in Table 5.4. Table 5.5 compares the gain performance obtained from the present method and those optimized by BBO [36] and CLPSO [35]. It clearly indicates the superior performance of the optimized loaded Yagi-Uda antenna over those unloaded BBO and CLPSO optimized results. Owing to the loading technique, an increased gain of 1.71 dB is obtained. However, as in the BBO and CLPSO optimization cases, the input impedance of the optimal gain antenna has not been matched to the standard $50\,\Omega$ transmission line. The radiation pattern of the Yagi-Uda Antenna I in the H-plane is shown in Figure 5.11. As can be seen, the optimized gain is about 15.55 dB, and the SLL is around -9.82 dB. The reactively loaded Yagi-Uda antenna will be further optimized to improve its input impedance matching property.

TABLE 5.4   Design Parameters of the Two Six-element Reactively Loaded Yagi-Uda Antennas

	Yagi-Uda antenna I			Yagi-Uda antenna II		
Units	λ		Ω	λ		Ω
Elements	Length, L_n	Spacing, S_n	Loading, X_{Ln}	Length, L_n	Spacing, S_n	Loading, X_{Ln}
1	0.24	0.17	25.5	0.24	0.22	−1.9
2	0.23	0.22	—	0.23	0.18	—
3	0.22	0.44	−23.0	0.22	0.27	9.2
4	0.21	0.40	−22.0	0.22	0.35	−14.8
5	0.21	0.40	−17.1	0.21	0.37	−18.0
6	0.21	—	0.5	0.20	—	58.2

TABLE 5.5   Performance Comparisons Between the Six-element Reactively Loaded Yagi-Uda Antenna and the Optimized Non-loaded Yagi-Uda Antennas

	Gain optimized			Gain and input impedance optimized		
	Antenna I	BBO [36]	CLPSO [35]	Antenna II	BBO [36]	CLPSO [35]
Gain (dBi)	15.55	13.84	13.84	13.44	12.69	12.65
Z_{in} (Ω)	16.2−j3.1	3.2+j23.4	3.9+j24.2	51.0−j0.1	50.0−j0.01	50.0−j0.01

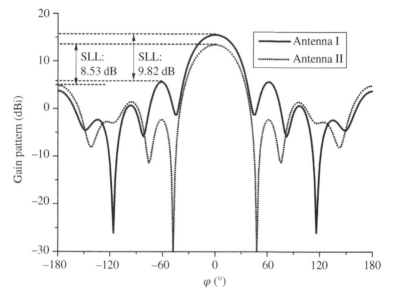

FIGURE 5.11   Radiation patterns of the reactively loaded six-element Yagi-Uda antenna.

As the second design case, we intentionally alter the physical dimensions of the Yagi-Uda antenna to show whether the physical dimensions are critical to the radiation performance. We refer to this antenna as the Yagi-Uda Antenna II. The element lengths and element spacings are also given in Table 5.4. In this design, we intend to maximize the gain and the input impedance simultaneously.

The optimized reactance loadings for Yagi-Uda Antenna II are given in Table 5.4. The corresponding antenna gain and input impedance for the optimized reactance loadings are given in Table 5.5. They are also compared with those obtained by directly applying the optimization methods BBO [36] and CLPSO [35] to the element lengths and spacings. It shows that the antenna gain obtained from the CM-based approach is higher than those optimized unloaded Yagi-Uda antennas. An increased gain of 0.75 and 0.79 dB is achieved as compared to the non-loaded Yagi-Uda antenna. Meanwhile, the input impedance obtained from the reactively loaded antenna provides perfect input impedance matching for standard 50 Ω RF systems.

The H-plane gain pattern of the reactively loaded antenna is shown in Figure 5.11. As can be seen, the maximum SLL of Antenna II is −8.53 dB. The six-element

Yagi-Uda antennas we have optimized have well demonstrated that Yagi-Uda antenna performances can be improved by using the loading technique, other than only by optimizing the element lengths and spacings as in the conventional designs.

5.4.2.2 15-Element Yagi-Uda Antenna Designs In order to demonstrate that the CM-based approach is capable of synthesizing Yagi-Uda antennas with larger number of elements, a 15-element Yagi-Uda antenna is considered in this section. Similarly, element lengths and spacings are determined prior to the optimization procedure. The physical dimensions for the 15-element Yagi-Uda antennas are shown in Table 5.6.

Firstly, we intend to maximize the antenna gain and constrain the input impedance to $50\,\Omega$. This antenna is referred to as Yagi-Uda Antenna III in the following discussions. The fourth column of Table 5.6 gives the optimal reactance values for this design. Table 5.7 compares the performances of the Yagi-Uda Antenna III with those optimized unloaded antenna. As can be seen, the antenna gain we obtained is 0.35, 0.09, and 1.34 dB higher than those optimized by the CLPSO [35], CI [34], GA [32] methods, respectively. Although the increases in gain is rather limited, they are still quite attractive, because the gain increases obtained by a more powerful optimization technique is much less than what we have achieved using the loading technique.

Moreover, the input impedance of the optimized reactively loaded antenna precisely equals to $50\,\Omega$, which indicates a perfect impedance matching to the standard $50\,\Omega$ RF systems. However, the CLPSO [35], CI [34], GA [32] results deviate slightly

TABLE 5.6 Design Parameters of the Two 15-element Reactively Loaded Yagi-Uda Antennas

	Physical dimensions		Yagi-Uda antenna III	Yagi-Uda antenna IV
Units	λ		Ω	
Elements	Length, L_n	Spacing, S_n	Loading, X_{Ln}	Loading, X_{Ln}
1	0.24	0.17	18.0	−7.7
2	0.23	0.17	—	—
3	0.22	0.26	−8.3	0.28
4	0.22	0.31	2.8	−6.5
5	0.21	0.22	−20.7	−15.5
6	0.20	0.26	−260.3	10.0
7	0.21	0.38	−25.6	5.7
8	0.20	0.34	−5.7	16.1
9	0.20	0.38	−3.4	−9.9
10	0.20	0.32	26.7	−9.3
11	0.20	0.41	26.1	4.6
12	0.20	0.21	6.8	−82.1
13	0.20	0.33	3.7	−129.4
14	0.20	0.37	−32.1	−56.6
15	0.20	—	8.21	9.8

TABLE 5.7 **Performance Comparison Between the Reactively Loaded 15-element Yagi-Uda Antenna and the Optimized Non-loaded Yagi-Uda Antennas**

	Designs using CMs		Designs in references		
	Antenna III	Antenna IV	CLPSO [35]	CI [34]	GA [32]
Gain (dB)	16.75	15.56	16.40	16.66	15.41
SLL (dB)	−10.04	−18.17	—	—	—
Z_{in} (Ω)	50.0+j0.0	50.03+j0.003	50.09+j0.15	45.42−j5.74	50.64−j5.08

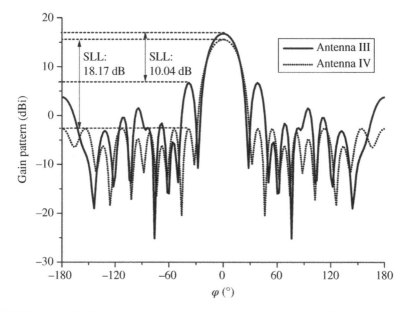

FIGURE 5.12   Radiation patterns of the reactively loaded 15-element Yagi-Uda antennas.

from the standard 50Ω impedance. Figure 5.12 shows the radiation patterns of the reactively loaded 15-element Yagi-Uda antennas. As can be seen, the maximum SLL is only −10.04 dB, which may be too large for some particular applications.

Therefore, in order to suppress the maximum SLL, we further optimize the 15-element Yagi-Uda antenna. In this new design (we refer it to Yagi-Uda Antenna IV), the antenna gain, SLL, and input impedance are optimized simultaneously by invoking the cost function in Equation (5.27). The optimized reactance loadings for Yagi-Uda Antenna IV are shown in the last column of the Table 5.6. Table 5.7 also shows the antenna gain, the SLL, and the input impedance of the Yagi-Uda Antenna IV. The radiation pattern of the Yagi-Uda Antenna IV is given in Figure 5.12. As can be seen, the maximum SLL has been suppressed to −18.17 dB, that is, an 8.13 dB SLL reduction is achieved. As compared to the Yagi-Uda Antenna III, the cost for the SLL reduction is that the antenna gain is reduced from 16.75 to 15.56 dB. The input impedance also matches well to the standard 50 Ω RF systems. As can be seen, there is a tradeoff in the pursuing of high gain and low SLL in Yagi-Uda antennas. The input impedance is not affected much in the SLL reduction. This example also demonstrates that the SLL of a Yagi-Uda antenna can be suppressed efficiently by loading the parasitic elements.

5.4.3 Efficiency Investigation

We have shown that the design method based on the CM theory for *N*-port networks is capable of optimizing the antenna gain, SLLs, and input impedance of reactively loaded Yagi-Uda antennas. The reactively loaded antennas usually have better performance than their unloaded counterparts. This section investigates the synthesis efficiency of the CM-based approach in the Yagi-Uda antenna designs.

In the optimization theory, the number of the cost function evaluations for a given convergence threshold is generally used to assess the efficiency of an optimization procedure. We also adopt this scheme to demonstrate the efficiency of the CM-based approach. Figure 5.13 shows the convergence rates of the cost function in four Yagi-Uda antenna designs. As can be seen, the CM-based approach requires approximately 10,000 and 50,000 cost function evaluations for the six-element and 15-element design cases, respectively.

It is well known that in electromagnetic optimization problems, the cost function evaluations are the most time-consuming part in the entire optimization procedure. The improvements to the optimization algorithm itself do not contribute much to the optimization efficiency enhancement. Owing to the orthogonality properties in the port currents and far-field radiations, the modal solutions in the CM theory replace the traditional full-wave analysis with a simple linear superposition of various characteristic modes. The computation time involved in the cost function evaluations is greatly reduced. Meanwhile, the accuracy of the candidate solution assessment keeps the same as that of full-wave simulations.

Table 5.8 compares the CPU time for 10,000 and 50,000 candidate solution evaluations in the six-element and 15-element Yagi-Uda antenna design cases. All these computations are conducted on a personal computer with one 3.40 GHz Intel® Core and 8 GB RAM. As can be seen, the CPU time reductions in the CM-based approach are quite evident. Therefore, although nearly 50,000 cost function evaluations are required in the 15-element Yagi-Uda antenna optimizations, the CPU time consumed in the whole design is really very short.

In summary, the CM theory is combined with the DE algorithm for the optimal design of reactively loaded Yagi-Uda antennas. Based on the characteristic mode analysis, the DE algorithm is applied to find the optimal modal expansion coefficients for each mode. The reactance values for each parasitic element can be calculated directly from the synthesized port currents. In this approach, the characteristic modes are very important decisions for the final designs. The characteristic modes also significantly enhance the efficiency of the whole synthesis procedure.

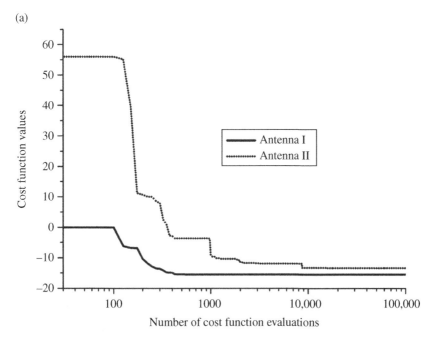

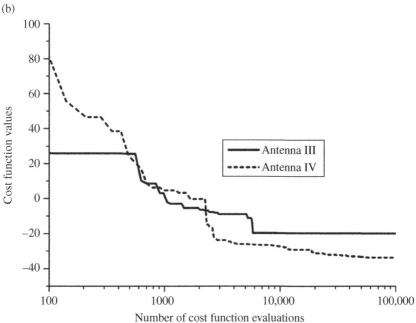

FIGURE 5.13 Convergence rates of the CM-based optimal designs of Yagi-Uda antennas. (a) Convergence rates in the six-element antenna designs and (b) convergence rates in the 15-element antenna designs.

TABLE 5.8   Cost Function Evaluation CPU Time in the CM-based Approach and the MoM-based Approach

Number of objective function evaluations	CPU time (s)	
	Characteristic modes	Method of moments
Six-element Yagi-Uda: 10,000	0.0624	611.2
Fifteen-element Yagi-Uda: 50,000	1.09	46,865.82

5.5   WIDEBAND ARRAY EXCITATION DESIGN USING CHARACTERISTIC MODES

5.5.1   Tightly Coupled Wideband Arrays

Wideband antenna arrays play an indispensable role in the emerging multifunctional RF systems, advanced radar systems, and high-resolution imaging systems. Wideband antenna arrays with a low-profile radiating aperture pose critical challenges for traditional antenna array theory. In the past decade, tightly coupled array has emerged as a very attractive candidate for wideband array designs [38–43]. It provides large bandwidths (over 4:1 bandwidth ratio) with very low-profile configurations. The thickness of the array is only around $\lambda/10$ at the lowest operation frequency.

The design concept of the tightly coupled wideband arrays is different from conventional wideband arrays. There are three most distinct features in tightly coupled wideband arrays:

1. The antenna elements are electrically small and they are closely spaced with each other. Both of the element sizes and element spacings are around $\lambda/10$ at the lowest operation frequency.
2. Strong capacitive couplings are introduced between adjacent elements. The two tips of the interdigital capacitors directly connect the adjacent elements or the adjacent elements are connected or overlapped.
3. One or more wide-angle impedance matching (WAIM) layers are usually implemented to compensate the impedance variance due to beam scanning.

The first two features ensure the antenna elements are tightly coupled. The strong capacitive couplings counteract the ground plane inductance and keep the input impedance be quasi-real over a wide bandwidth. It enables 4:1 impedance bandwidth or even much wider bandwidth [43–45].

Because of the strong mutual coupling between antenna elements, uniform element excitation no longer provides maximum aperture efficiency. Moreover, uniform element excitation causes serious mismatching issues in the edge elements [9]. In order to achieve quasi-optimal array excitations for a finite tightly coupled antenna array, a novel approach based on the characteristic mode theory for N-port networks is developed. Unlike the CM-based radiation pattern synthesis approaches in the preceding two sections, this CM-based approach aims to provide wideband matching for all of the array elements, including the edge elements in a finite array. It allows the array having maximum aperture efficiency over an ultra-wide frequency range, say around 4:1 bandwidth.

5.5.2 Computation of Quasi-Excitations from Characteristic Modes

As discussed in Section 5.2, the characteristic modes of an antenna array can be solved from the generalized eigenvalue equation defined for the N-port impedance matrix Z. Following the CM theory, maximum aperture efficiency can be achieved if the characteristic current $\bar{J}_n$ with zero eigenvalue ($\lambda_n = 0$) is excited at the resonant frequency. Because the modal current $\bar{J}_n$ is a real vector, the port voltage $\bar{V}_n$ corresponding to the characteristic currents $\bar{J}_n$ with $\lambda_n = 0$ can be written as follows by considering the generalized eigenvalue equation $[X]\bar{J}_n = \lambda_n [R]\bar{J}_n$ [9]:

$$
\begin{aligned}
\bar{V}_n &= [Z]\bar{J}_n \\
&= [R]\bar{J}_n + j[X]\bar{J}_n \\
&= [R]\bar{J}_n + j\lambda_n [R]\bar{J}_n \\
&\approx [R]\bar{J}_n
\end{aligned}
\tag{5.36}
$$

Based on Equation (5.36), the active input impedance $\bar{Z}_a$ at the excitation ports of an antenna array can be found from:

$$
\bar{Z}_a = \bar{V}_n ./ \bar{I}_n \approx \{[R]\bar{J}_n\}./ \bar{J}_n
\tag{5.37}
$$

where ./ denotes the element-wide division of two vectors. Because $[R]$ and $\bar{J}_n$ are real matrix and real vector, respectively, the active input impedances at each port are also real. It indicates that all the ports can be matched simultaneously if the characteristic impedance of the feeding transmission line is equal to the active input impedance.

On the other hand, the S-parameters are more popular in the assessment of the matching performance, and the excitation coefficients for the incident power waves of each port are required in practical antenna array excitation. The standard definitions for the S-parameters show that the incident and reflected waves are related to the input impedance of feeding lines by Ref. [46]:

$$
[Z_0]^{-1/2} \bar{V}_n = \bar{a} + \bar{b}
\tag{5.38}
$$

$$
[Z_0]^{1/2} \bar{I}_n = \bar{a} - \bar{b}
\tag{5.39}
$$

where $[Z_0]$ is a diagonal matrix containing the characteristic impedances of the feeding lines. $\bar{V}_n$ and $\bar{J}_n$ are the port voltages and port currents of the resonant characteristic mode with $\lambda_n = 0$. $\bar{a}$ and $\bar{b}$ consist of the coefficients for the incident and reflected waves at the array ports, respectively. By adding Equations (5.38) and (5.39), we can get the excitation coefficients for the resonant characteristic mode:

$$\bar{a} = \frac{1}{2}\left([Z_0]^{-1/2}\,\bar{V}_n + [Z_0]^{1/2}\,\bar{I}_n\right) \tag{5.40}$$

Meanwhile, the diagonal elements $Z_{0i}\ (i = 1,...,N)$ in $[Z_0]$ can be calculated from the following:

$$Z_{0i} = \left[\bar{V}_n./\bar{I}_n\right]_i \tag{5.41}$$

To this end, the array excitation taper $\bar{a}$ and the feeding line impedance $Z_{0i}\ (i = 1,...,N)$ are obtained. Because they are computed from the characteristic mode at the exact resonant frequency, the array excitations will give the maximum radiation efficiency if each of the port is fed by a transmission line with a characteristic impedance of Z_{0i}.

The above array excitations are solved from the characteristic mode at a single resonant frequency. Therefore, they achieve perfect impedance matching at this resonant frequency. However, the goal in wideband array designs is to match the feeding ports over a wide frequency range. This problem was discussed in detail through an 8×8 tightly coupled planar antenna array in Ref. [9]. In Ref. [9], the 8×8 planar array is taken as a 64-port network. The characteristic modes are solved from the 64×64 mutual impedance matrix of the planar array. The mutual impedance matrix of the antenna array can be either measured from a vector network analyzer or calculated using various electromagnetic numerical techniques. In the present work, the mutual impedance matrix was solved from the HFSS software [46]. By solving the generalized eigenvalue equation in Equation (5.6), the modal significance of the planar array can be obtained. As a rule of thumb, characteristic modes with modal significance larger than 0.6 can be recognized as significant modes that are efficient for radiation purpose. The numerical results demonstrate that by relaxing the resonance condition from $MS = 1$ to $MS = 0.6$, it is possible to achieve satisfactory impedance matching over a wide frequency range with the CM-based excitation taper.

As a conclusion, the quasi-optimal array excitation calculated from Equation (5.40) for achieving optimal impedance matching in all array elements can be derived from the characteristic port currents of the tightly coupled antenna array. In particular, the CM-based excitation technique is more suitable to antenna arrays with strong mutual couplings. For antenna arrays with weakly coupled elements, the mutual impedance matrix is nearly a diagonal matrix. The impedance matching in each antenna element is dominated by the element itself, other than the mutual couplings. Therefore, there will be no significant differences in the impedance matching performance of the CM-based excitations and uniform excitations. The CM-based technique can be also applied to antenna arrays with measured mutual impedance matrix. It is particularly attractive in the design of high-power antenna arrays, where the simultaneous impedance matching for all the elements is of great importance.

5.6 CONCLUSIONS

In this chapter, we described the characteristic mode theory for N-port networks. The characteristic modes form a convenient basis for expressing the radiating field and port currents of an N-port network. The characteristic fields and characteristic currents hold the orthogonality over the radiation sphere and the ports, respectively. These attractive features found wide applications in the analysis, synthesis, and optimization designs of N-port networks.

In particular, the applications of the characteristic mode theory for N-port networks are illustrated through many antenna array designs. The design examples show that this CM theory is very attractive in the following aspects:

- For radiation pattern synthesis of reactively loaded multiport antenna systems, the design specifications may simultaneously include the main beam direction, the SLL, and the directivity.
- The modal solutions in the CM theory greatly enhance the synthesis efficiency, whereas the accuracy is kept the same as those in full-wave simulations.
- CM theory can determine the quasi-optimal array excitations from the characteristic currents. It ensures simultaneous impedance matching of all the elements over a wide frequency band.

In all of the design cases, either the characteristic fields or the characteristic currents explicitly determine the optimal design parameters. It avoids time-consuming full-wave simulations as in traditional antenna array design approaches. It also provides clear physical information to the operational principle of various multiport antenna systems.

REFERENCES

[1] J. Mautz and R. Harrington, "Modal analysis of loaded N-port scatterers," *IEEE Trans. Antennas Propag.*, vol. 21, no. 2, pp. 188–199, Mar. 1973.

[2] J. Ethier and D. McNamara, "An interpretation of mode-decoupled MIMO antennas in terms of characteristic port modes," *IEEE Trans. Magn.*, vol. 45, no. 3, pp. 1128–1131, Mar. 2009.

[3] J. Ethier, E. Lanoue, and D. McNamara, "MIMO handheld antenna design approach using characteristic mode concepts," *Microw. Opt. Technol. Lett.*, vol. 50, no. 7, pp. 1724–1727, Jul. 2008.

[4] K. A. Obeidat, B. D. Raines, and R. G. Rojas, "Application of characteristic modes and non-Foster multiport loading to the design of broadband antennas," *IEEE Trans. Antennas Propag.*, vol. 58, no. 1, pp. 203–207, Jan. 2010.

[5] R. F. Harrington, "Reactively controlled directive arrays," *IEEE Trans. Antennas Propag.*, vol. AP-26, no. 3, pp. 390–395, May 1978.

[6] W. X. Zhang and Y.-M. Bo, "Pattern synthesis for linear equal-spaced antenna array using an iterative eigenmodes method," *IEE Proc. H, Microw. Antennas Propag.*, vol. 135, no. 3, pp. 167–170, Jun. 1988.

[7] Y. Chen and C.-F. Wang, "Synthesis of reactively controlled antenna arrays using characteristic modes and DE algorithm," *IEEE Antennas Wirel. Propag. Lett.*, vol. 11, pp. 385–388, 2012.

[8] Y. Chen and C.-F. Wang, "Electrically loaded Yagi-Uda antenna optimizations using characteristic modes and differential evolution," *J. Electromagn. Waves Appl.*, vol. 26, pp. 1018–1028, 2012.

[9] I. Tzanidis, K. Sertel, and J. L. Volakis, "Characteristic excitation taper for ultrawide-band tightly coupled antenna arrays," *IEEE Trans. Antennas Propag.*, vol. 60, no. 4, pp. 1777–1784, Apr. 2012.

[10] R. Harrington and J. Mautz, "Pattern synthesis for loaded *N*-port scatterers," *IEEE Trans. Antennas Propag.*, vol. 22, no. 2, pp. 184–190, Mar. 1974.

[11] R. F. Harrington and J. R. Mautz, "Computation of characteristic modes for conducting bodies," *IEEE Trans. Antennas Propag.*, vol. AP-19, no. 5, pp. 629–639, Sep. 1971.

[12] S. Chen, A. Hirata, T. Ohira, and N. C. Karmakar, "Fast beamforming of electronically steerable parasitic array radiator antennas: theory and experiment," *IEEE Trans. Antennas Propag.*, vol. 52, no. 7, pp. 1819–1832, Jul. 2004.

[13] R. F. Harrington and J. R. Mautz, "Reactively loaded directive arrays," Rep. TR-74-6, Contract N00014-67-A-0378-0006 between the Office of Naval Research and Syracuse Univ., Syracuse, NY, Sep. 1974.

[14] H. Scott and V. F. Fusco, "360 electronically controlled beam scan array," *IEEE Trans. Antennas Propag.*, vol. 52, no. 1, pp. 333–335, Jan. 2004.

[15] S. Sugiura and H. Iizuka, "Reactively steered ring antenna array for automotive application," *IEEE Trans. Antennas Propag.*, vol. 55, no. 7, pp. 1902–1908, Jul. 2007.

[16] H. Kato and Y. Kuwahara, "Novel ESPAR antenna," in Proceedings of the Antennas and Propagation Society International Symposium, Washington, DC, vol. 4B, pp. 23–26, Jul. 2005.

[17] Y. Chen, S. Yang, and Z. Nie, "Synthesis of uniform amplitude thinned linear phased arrays using the differential evolution algorithm," *Electromagnetics*, vol. 27, no. 5, pp. 287–297, Jun.–Jul. 2007.

[18] R. F. Harrington, *Field Computation by Moment Methods*. New York: Macmillan, 1968.

[19] P. P. Silvester and R. L. Ferrari, *Finite Elements for Electrical Engineers* (3rd edition). Cambridge, UK: Cambridge University Press, 1996.

[20] J.-M. Jin, *The Finite Element Method in Electromagnetics* (2nd edition). New York: John Wiley & Sons, Inc., 2002.

[21] R. L. Haupt, "Optimized element spacing for low sidelobe concentric ring arrays," *IEEE Trans. Antennas Propag.*, vol. 56, no. 1, pp. 266–268, Jan. 2008.

[22] B. Fuchs, R. Golubovic, A. K. Skrivervik, and J. R. Mosig, "Spherical lens antenna designs with particle swarm optimization," *Microw. Opt. Technol. Lett.*, vol. 52, no. 7, pp. 1655–1659, Jul. 2010.

[23] A. Qing, *Differential Evolution: Fundamentals and Applications in Electrical Engineering*. New York, NY: John Wiley & Sons, Inc., 2009.

[24] D. Mandal, S. P. Ghoshal, and A. K. Bhattacharjee, "Radiation pattern optimization for concentric circular antenna array with central element feeding using craziness-based particle swarm optimization," *Int. J. RF Microwave Comput. Aided Eng.*, vol. 20, no. 5, pp. 577–586, Sep. 2010.

[25] H. Yagi and S. Uda, "Projector of the sharpest beam of electric waves," *Proc. Imp. Acad. Jpn.*, vol. 2, no. 2, pp. 49–52, Feb. 1926.

[26] C. A. Balanis. *Modern Antenna Handbook*. Hoboken, NJ: John Wiley & Sons, Inc., 2008.

[27] H. W. Ehrenspeck and H. Poehler, "A new method for obtaining maximum gain from Yagi antennas," *IRE Trans. Antennas Propag.*, vol. 7, pp. 379–386, 1959.

[28] D. K. Cheng and C. A. Cheng, "Optimum element spacing for Yagi-Uda arrays," *IEEE Trans. Antennas Propag.*, vol. 21, pp. 615–623, 1973.

[29] D. K. Cheng and C. A. Cheng, "Optimum element length for Yagi-Uda arrays," *IEEE Trans. Antennas Propag.*, vol. 23, pp. 8–15, 1975.

[30] D. K. Cheng, "Gain optimization for Yagi-Uda arrays," *IEEE Antennas Propag. Mag.*, vol. 33, pp. 42–45, 1991.

[31] E. E. Altshuler and D. S. Linden, "Wire-antenna designs using genetic algorithms," *IEEE Antennas Propag. Mag.*, vol. 39, no. 2, pp. 33–43, 1997.

[32] E. A. Jones and W. T. Joines, "Design of Yagi-Uda antennas using genetic algorithms," *IEEE Trans. Antennas Propag.*, vol. 45, pp. 1386–1392, Sep. 1997.

[33] J. D. Lohn, W. F. Kraus, and S. P. Colombano, "Evolutionary optimization of Yagi-Uda antennas," in Proceedings of the 4th International Conference Evolvable Systems, Tokyo, Japan, pp. 236–243, Oct. 3–5, 2001.

[34] N. V. Venkatarayalu and T. Ray, "Optimum design of Yagi-Uda antennas using computational intelligence," *IEEE Trans. Antennas Propag.*, vol. 52, no. 7, pp. 1811–1818, Jul. 2004.

[35] S. Baskar, A. Alphones, P. N. Suganthan, and J. J. Liang, "Design of Yagi-Uda antennas using comprehensive learning particle swarm optimization," *IEE Proc. Microw. Antennas Propag.*, vol. 152, no. 5, pp. 340–346, Oct. 2005.

[36] U. Singh, M. Rattan, and N. Singh, "Optimization of Yagi antenna for gain and impedance using simulated annealing," in Proceedings of the IEEE, ICECOM, Dubrovnik, Croatia, vol. 24–26, pp. 1–4, Sep. 2007.

[37] U. Singh, H. Kumar, and T. S. Kamal, "Design of Yagi-Uda antenna using biogeography based optimization," *IEEE Trans. Antennas Propag.*, vol. 58, no. 10, pp. 3375–3379, Oct. 2010.

[38] B. Munk, "Chapter 6," in *Finite Antenna Arrays and FSS*. Hoboken, NJ: John Wiley & Sons, Inc., 2003.

[39] B. Munk, R. Taylor, T. Durharn, et al., "A low-profile broadband phased array antenna," in Proceedings of the IEEE S-AP International Symposium, Columbus, OH, vol. 2, pp. 448–451, Jun. 9–11, 2003.

[40] M. Jones and J. Rawnick, "A new approach to broadband array design using tightly coupled elements," in Proceedings of the IEEE MILCOM, Orlando, FL, pp. 1–7, Oct. 29–31, 2007.

[41] E. O. Farhat, K. Z. Adami1, Y. Zhang, A. K. Brown, et al., "Ultra-wideband tightly coupled phased array antenna for low-frequency radio telescope," in Progress In Electromagnetics Research Symposium Proceedings, Stockholm, Sweden, pp. 245–249, Aug. 12–15, 2013.

[42] Y. Chen, S. Yang, and Z. Nie, "A novel wideband antenna array with tightly coupled octagonal ring elements," *Prog. Electromagn. Res.*, vol. 124, pp. 55–70, 2012.

[43] Y. Chen, S. Yang, and Z. Nie, "The role of ground plane plays in wideband phased array antenna," in 2010 IEEE International Conference on Ultra-Wideband, ICUWB, 2010-Proceedings, Nanjing, vol. 2, pp. 829–831, Sep. 2010.

[44] J. M. Bell, M. F. Iskander, and J. J. Lee, "Ultrawideband hybrid EBG/ferrite ground plane for low-profile array," *IEEE Trans. Antennas Propag.*, vol. 55, no. 1, pp. 4–12, Jan. 2007.

[45] J. J. Lee, S. Livingston, and D. Nagata, "A low profile 10:1 (200–2000 MHz) wide band long slot array," in IEEE International Symposium on Antennas and Propagation, San Diego, CA, pp. 1–4, Jul. 2008.

[46] ANSYS Corporation [Online]: http://www.ansys.com/Products/Simulation+Technology/ Electronics/Signal+Integrity/ANSYS+HFSS. Accessed January 28, 2015.

6

PLATFORM-INTEGRATED ANTENNA SYSTEM DESIGN USING CHARACTERISTIC MODES

6.1 BACKGROUNDS

As have been discussed in Chapters 2 through 5, characteristic mode (CM) theory could be developed for many different electromagnetic structures. In addition, we have also demonstrated that these CMs can provide in-depth physical insights and valuable design guidelines for a variety of antennas, including the metallic antennas, microstrip patch antennas, dielectric resonator antennas, multiport antennas, and antenna arrays. Based on the CM theory for perfectly electrically conducting (PEC) bodies, this chapter goes on to discuss the applications of the CM theory in a new class of antenna design problems: the platform-integrated antenna system designs.

Designing an antenna with required platform-installed performance is challenging and has been investigated by many researchers. Traditionally, the design process begins with developing an isolated antenna with satisfactory radiation performance. Following such an optimal design, the isolated antenna is then installed on a mounting platform and is expected to work well within the platform environment. Unfortunately, the couplings from the platform may lead to severe deteriorations in the radiation performances of the antenna. Such a situation is frequently encountered in high frequency (HF)/very high frequency (VHF) antennas mounted on platforms with only a few wavelengths [1–3]. In some cases, these adverse effects can be alleviated by judiciously choosing the antenna installation location. However, it does not work for all the cases.

Characteristic Modes: Theory and Applications in Antenna Engineering, First Edition.
Yikai Chen and Chao-Fu Wang.
© 2015 John Wiley & Sons, Inc. Published 2015 by John Wiley & Sons, Inc.

A general approach is to co-optimize the antenna with the consideration of the platform. In the past decades, various electromagnetic numerical techniques have been developed to analyze the radiation performances of platform-installed antenna systems. The co-optimization of antennas with platform is possible by adopting these electromagnetic numerical techniques. However, these techniques are only capable of evaluating the radiation performance of an existing design. They cannot provide clear physical insights to our antenna engineers for further optimizations.

Evolutionary algorithms are also extensively used to accommodate the co-optimization design of antennas in platform environment. For instance, the finite difference time domain (FDTD) method [4, 5] was combined with the genetic algorithm to seek for the optimal antenna locations [6–8]. However, determining the fitness of each candidate solution involves extremely long time in the full-wave simulation. Meanwhile, evolutionary algorithms usually require thousands of fitness evaluations to ensure the convergence to a near-optimal solution. As such, such design tool can be extremely computationally expensive. In order to overcome such limitations, graphic processor unit (GPU) has been adopted to speed up the FDTD analysis [7].

As can be seen, despite the advancements in electromagnetic numerical techniques and evolutionary optimization algorithms, optimal design of platform-installed antennas remains a big challenge. Therefore, there is a need to develop an effective and versatile tool for platform integrated antenna system designs, especially for antenna systems mounted on electrically small or moderate platforms in the HF/VHF frequency range.

In recent years, there are many attempts to address the platform-installed antenna system designs by using the CM theory [9–17]. For instance, the optimal location of a small loop antenna on a simple wire cross supporting structure was determined from the CMs of the supporting structure [9]. Theoretical study showed that maximum radiation efficiency can be achieved by placing the loop antenna at the current maximum of the dominant CM. In Refs. [12, 13], near-vertical incidence skywave (NVIS) patterns were synthesized from the CMs of the entire structure consisting of a land vehicle and a large loop antenna. Ideal voltage sources and reactive loadings are placed on the big loop to excite the current for the NVIS pattern. The possibility to design conformal antennas following the CMs of an electrically small unmanned aerial vehicle (UAV) has been investigated in Ref. [10]. These early works illustrate that the CM theory may provide a new perspective to investigate the platform-integrated antenna system designs. Although practical excitation method and input impedance matching issue have not been well addressed in the early works, it has showed that the radiation efficiency can be enhanced and the radiation pattern can be controlled by properly utilizing the CMs of the platform.

In this chapter, an advanced systematic procedure for platform-integrated antenna system designs using the CMs of the platforms is fully described [18]. Attractive features of this approach, including the clear physical meanings, the design efficiency (radiating pattern synthesis and feeding designs), the improved antenna performances (gain, efficiency, and impedance matching), and the low profile or even platform-embedded configurations, will be demonstrated through the following two design cases:

1. HF band electrically small UAV antenna system designs
2. HF band shipboard antenna system designs

Both the designs are experimentally verified in their corresponding scaled models. Practical implementations of these designs in full model are also discussed.

6.2 ELECTRICALLY SMALL UAV ANTENNA SYSTEM DESIGN USING CHARACTERISTIC MODES

This section discusses the electrically small UAV antenna system designs using the CM theory. With the knowledge of the CMs of the UAV body, a multiobjective evolutionary algorithm is implemented to synthesize currents on the UAV body for achieving desired power patterns. Compact and low-profile feed structures are then designed to excite the synthesized currents. The feed structures are termed as probes, as they are used to excite the currents on the UAV body and cannot work as stand-alone antennas. The UAV body excited by these probes serves as the radiating aperture. Its reconfigurable radiation patterns are obtained by feeding each probe with proper magnitude and phase. In this method, the aperture of the UAV body is fully utilized. Practical issues associated with large antennas at low-frequency band are eliminated. Knowledge for probe placement on such platform becomes more explicit. Simulated and measured results are presented to verify the design concept.

6.2.1 Reconfigurable Radiation Pattern Synthesis Using Characteristic Modes

With the purpose of obtaining a prototype with reasonable physical size for tailoring to the available frequency range of the microwave chamber, practical UAV working at the HF range is scaled down to the UHF range. In this section, an electrically small UAV with electrical size of $0.648\lambda \times 0.746\lambda \times 0.154\lambda$ as shown in Figure 6.1 is considered. The entire UAV body is used as the radiator for the synthesis of reconfigurable radiation patterns at 800 MHz. Without loss of generality, the efficiency and effectiveness of the proposed method will be discussed throughout this design.

6.2.1.1 Multiobjective Optimizer: MOEA/D Multiobjective optimization methods are quite attractive in antenna designs due to their multiobjective nature as reported in Refs. [19, 20]. Actually, the reconfigurable radiation pattern synthesis problem is also a multiobjective problem. The design objectives include the main beam directions, front-to-back ratio, and antenna gain. To balance these conflicting goals, an efficient multiobjective evolutionary algorithm based on decomposition (MOEA/D) [21, 22] is adopted in this design. We will demonstrate that the MOEA/D can be used

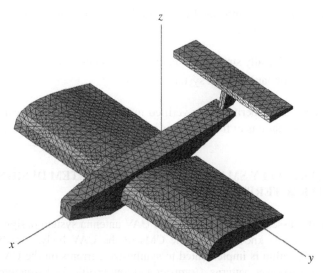

FIGURE 6.1   Geometry and mesh of an UAV used as the supporting structure.

to find many Pareto-optimal solutions in a single run, and the desired solution can be chosen manually following the design preferences.

A general definition for a multiobjective optimization problem with m objectives can be mathematically expressed as follows:

$$\text{minimize } F(\mathbf{x}) = \{f_i(\mathbf{x}), i = 1, 2, \ldots, m\}, \mathbf{x} \in \Omega \tag{6.1}$$

where Ω is the decision (variable) space, $F : \Omega \to R^m$ consists of m real-valued objective functions $f_i(\mathbf{x})$, and R^m is called the objective space. $\mathbf{x}$ is a general representation of a potential solution. Taking the platform antenna synthesis problem as an example, R^m is the objective space with design goals such as the main beam direction, front-to-back ratio, and antenna gain, whereas Ω is a decision space with the modal expansion coefficients a_n in Equations (2.75)–(2.77). It is quite common to encounter problems with conflicting objectives; therefore, one has to balance them, and the goal of multiobjective optimization is to find a Pareto-optimal set. The term "Pareto-optimal" means there is no absolute global optimal solution to the problem and there is no solution better than any other solution.

To find the Pareto-optimal solution in Equation (6.1), MOEA/D begins by decomposing the Pareto front into a number of scalar optimization problems. In this work, Tchebycheff decomposition approach is applied, and the scalar optimization problems can be written as follows:

$$\text{minimize } g^{\text{te}}(\mathbf{x} \mid \boldsymbol{\chi}, z^*) = \max_{i \in [1, m]}\left\{ \chi_i \mid f_i(\mathbf{x}) - z_i^* \right\} \tag{6.2}$$

where $\boldsymbol{\chi} = (\chi_1, \ldots, \chi_m)$ is a weight vector to aggregate the objective functions in Equation (6.1). $z^* = \{z_i^*, i \in [1, m]\}$ is defined as the reference point, and each element in the reference point has the minimum objective function value in its decision space, that is, $z_i^* = \min\{f_i(\mathbf{x}) \mid \mathbf{x} \in \Omega\}$ for each $i \in [1,m]$. $\mathbf{x}$ is a solution in the decision space Ω. In the evolutionary computation community, it is well known that for each Pareto-optimal point there is a weight vector χ such that it is the optimal solution of Equation (6.2). Each optimal solution of Equation (6.2) is one of the Pareto-optimal solutions of the original problem in Equation (6.1). This equivalence has been proved and is extensively applied in the development of various evolutionary optimization methods [21, 22]. Therefore, one can obtain the approximated Pareto front of the original problem by solving the N scalar sub-problems associated with different weight vector χ. General steps of the MOEA/D algorithm are summarized in Ref. [20].

6.2.1.2   CM Based Pattern Synthesis Method   Following the modal solutions formulated in Equations (2.75)–(2.77), pattern synthesis using the CMs of the platform is to seek for a set of modal coefficients such that the corresponding radiation fields **E** and **H** satisfy the specified requirements on the radiation pattern. Therefore, CM analysis for the UAV should be performed first. Figure 6.2 shows the eigenvalues over a frequency band. It is observed that the first dominant mode resonates at 800 MHz, and high-order modes resonate at much higher frequency band. At these resonant frequencies, the absolute eigenvalues of the resonant modes are close to zero (<−20 dB). It is also clearly observed that the absolute eigenvalues of high order modes increase rapidly upto 5 dB at the 800 MHz. Therefore, taking into account the contributions of the first few modes are sufficient in the synthesis of designated radiation patterns.

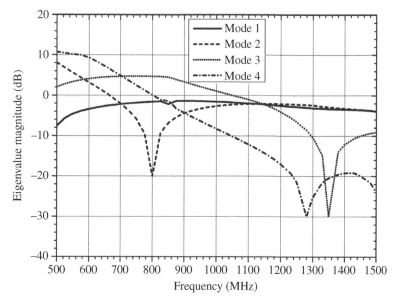

FIGURE 6.2   Eigenvalues of the first four modes for the UAV.

Figure 6.3 shows the characteristic currents and characteristic fields of the first four modes at 800 MHz. The frequency of interest for the radiation pattern synthesis is deliberately chosen at the resonant frequency of the dominant mode. One can also choose other frequencies for pattern synthesis. It can be seen clearly from Figure 6.3

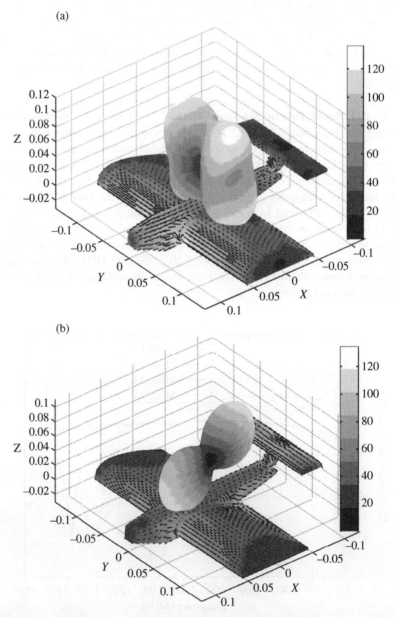

FIGURE 6.3 Characteristic currents and characteristic fields of the UAV at 800 MHz. (a) $\mathbf{J}_1$ & $\mathbf{E}_1$, (b) $\mathbf{J}_2$ & $\mathbf{E}_2$

(c)

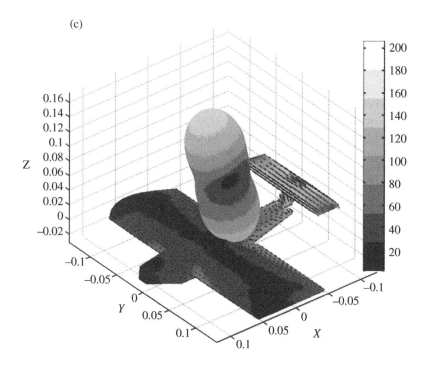

(d)

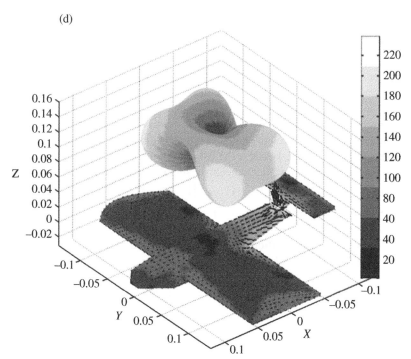

FIGURE 6.3   (*Continued*) (c) $\mathbf{J}_3$ & $\mathbf{E}_3$, and (d) $\mathbf{J}_4$ & $\mathbf{E}_4$.

that these modes are orthogonal with each other, in terms of both the currents and far fields. The characteristic fields of different modes have their main beams in different directions. Although none of the characteristic field pattern is consistent with typical radiation patterns like the directional radiation pattern or omnidirectional radiation pattern, they provide the design freedoms to synthesize any desired radiation patterns. Based on these modes, the MOEA/D is employed to seek for the optimal modal expansion coefficients for each mode.

Without loss of generality, we assume that directional radiation patterns with their main beams in the broadside, forward, and backward directions of the UAV fuselage are required for particular applications. Three objective functions are then constructed to guide the optimization,

$$f_1\left(\boldsymbol{\alpha}^{(k)}\right) = \left|\theta_0 - \theta_{\text{desired}}\right| + \left|\phi_0 - \phi_{\text{desired}}\right| \tag{6.3}$$

$$f_2\left(\boldsymbol{\alpha}^{(k)}\right) = -\frac{\oiint_{S_{\text{MR}}}(\mathbf{E}^* \times \mathbf{H}) \cdot d\mathbf{s}}{\oiint_{S_\infty}(\mathbf{E}^* \times \mathbf{H}) \cdot d\mathbf{s}} \tag{6.4}$$

$$f_3\left(\boldsymbol{\alpha}^{(k)}\right) = -\oiint_{S_{\text{MR}}}(\mathbf{E}^* \times \mathbf{H}) \cdot d\mathbf{s} \tag{6.5}$$

where θ_0 and φ_0 are the main beam directions corresponding to a particular set of modal coefficients $\boldsymbol{\alpha}^{(k)} = \left\{a_1^{(k)}, a_2^{(k)}, \ldots, a_N^{(k)}\right\}$ at the kth iteration in the multiobjective MOEA/D optimizer, and θ_{desired} and ϕ_{desired} are the desired main beam directions. N is the number of modes involved in the pattern synthesis. Numerical results prove that the first several low-order modes are sufficient to ensure the successful pattern synthesis. In Equation (6.3), $f_1(\boldsymbol{\alpha}^{(k)})$ constrains the directions of the main beams to the desired directions. The optimizer will iteratively minimize the angle errors. Equation (6.4) defines the ratio between the radiated power over the main beam range S_{MR} and the entire radiation sphere. Because the MOEA/D algorithm is developed to minimize the objective function values, a negative sign is assigned to the front of the ratio to maximize the term $\oiint_{S_{\text{MR}}}(\mathbf{E}^* \times \mathbf{H}) \cdot d\mathbf{s} / \oiint_{S_\infty}(\mathbf{E}^* \times \mathbf{H}) \cdot d\mathbf{s}$. $f_3(\boldsymbol{\alpha}^{(k)})$ is defined to constrain more energy in the main beam range. Figure 6.4 gives the flowchart of the pattern synthesis method. As can be seen, the MOEA/D optimizer continuously adjusts the modal coefficients until the objective function values converge to the preset threshold, and the antenna performance finally approaches to the desired one.

It is worth noting that far fields for any design candidates $\boldsymbol{\alpha}^{(k)}$ can be computed efficiently from Equations (2.76) and (2.77). Objective function values can then be computed from the linear combination of the CM fields. Because CMs with large eigenvalue magnitudes are inefficient radiating modes, only the first 10 CMs of the platform are used in radiation pattern synthesis. Because the modes with mode order

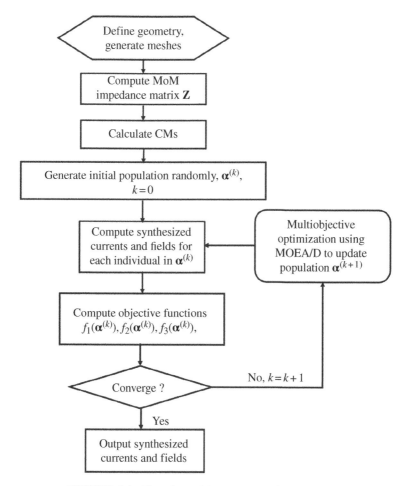

FIGURE 6.4 Flowchart of the pattern synthesis method.

higher than 10 usually have very large eigenvalue magnitude, and they cannot contribute much to the final synthesized radiation patterns, using more high-order modes in the synthesis makes no difference in the results. Because time-consuming full-wave simulations is eliminated, whereas the accuracy is kept the same as those obtained from full-wave simulations, more than three orders of speedup as compared to the traditional methods [6–8] is achieved.

6.2.1.3 Synthesis Results According to the requirements, we assume the pencil beams point to the broadside, forward, and backward directions with respect to the fuselage, the pattern synthesis method is employed to synthesize the current distributions for producing the aforementioned pencil beams. Objective functions defined in Equations (6.3)–(6.5) are considered in the MOEA/D optimizer. Figure 6.5 shows the final Pareto fronts of the solutions for the broadside,

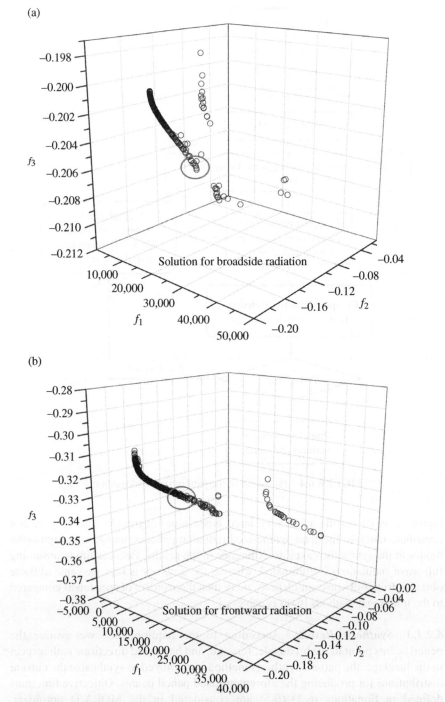

FIGURE 6.5 Pareto front of the errors for trading off the main beam directions, front-to-back ratios, and total energy within the main beam range. (a) Broadside radiation, (b) frontward radiation.

(c)

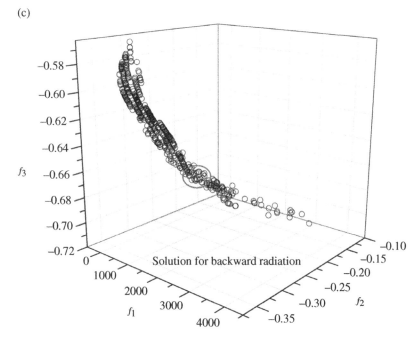

FIGURE 6.5   (*Continued*) (c) backward radiation.

forward, and backward cases. The coordinate values of the three axes represent the objective function values for $f_1(\alpha), f_2(\alpha)$, and $f_3(\alpha)$, respectively. Each point at the Pareto front represents a Pareto-optimal design, and no global optimal solution exists for all the objective functions. To make a tradeoff for the three design objectives, solutions within the grayed circle are chosen for further designs.

Figure 6.6 shows the synthesized currents and their corresponding radiation patterns from MOEA/D. It is observed that the broadside radiation is dominated by the currents on the wings and tail, whereas the forward and backward radiations are mainly contributed by the currents on the front edge of the wings and UAV tail, respectively. By comparing the real parts of the current distributions for the forward and backward cases, it is found that the current vectors almost have the same magnitudes and directions. However, the comparison between the imaginary parts of the current distributions for the forward and backward cases shows that they have the same current magnitude but opposite current vector directions. This observation illustrates that the currents for the forward and backward radiation have the same magnitudes but opposite phases. It is interesting to find that if we keep the current magnitude exactly the same as in Figure 6.6c and d, and make the current phases be opposite to the original one, the forward radiation pattern turns to backward radiation. Therefore, the aircraft current distributions can be simplified into something like strip currents. Actually, basic principles for phased arrays can be borrowed to interpret the above phenomenon. It can be predicted that if these currents could be excited in practice, we can get the reconfigurable radiation patterns. As we shall see, these current distributions

(a)

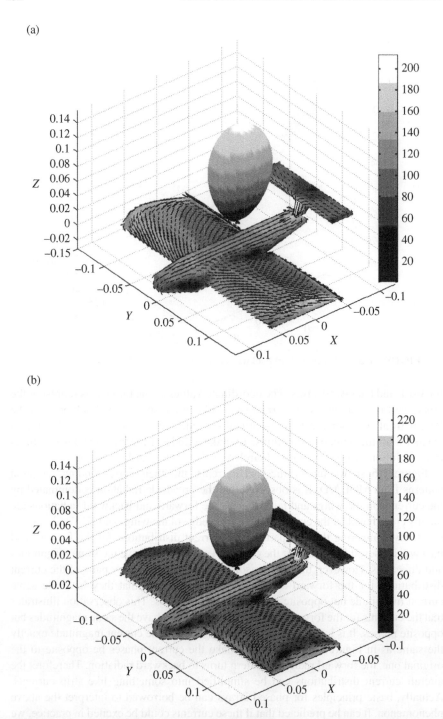

(b)

FIGURE 6.6   Synthesized currents and radiation patterns from MOEA/D. (a) Real part of the current for broadside radiation, (b) imaginary part of the current for broadside radiation.

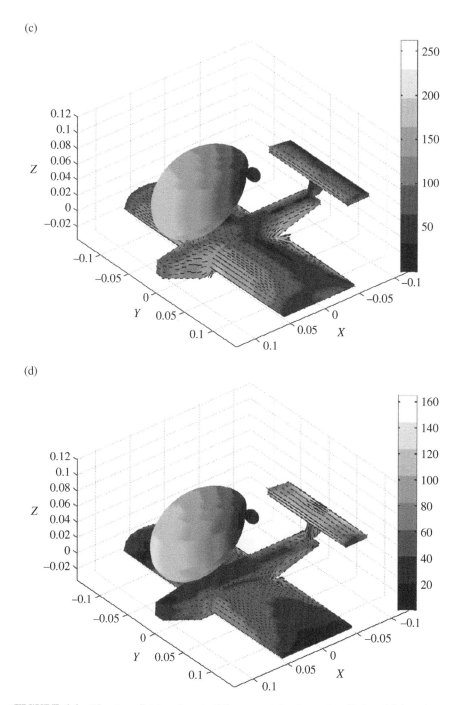

FIGURE 6.6 (*Continued*) (c) real part of the current for forward radiation, (d) imaginary part of the current for forward radiation.

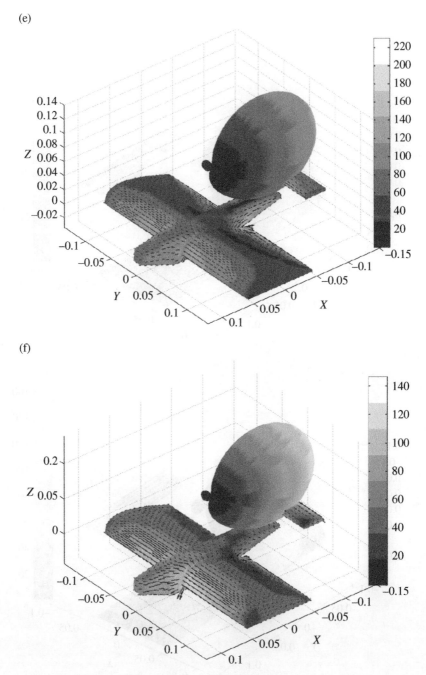

FIGURE 6.6 (*Continued*) (e) real part of the current for backward radiation, and (f) imaginary part of the current for backward radiation.

provide valuable information on how to design and place the feeding structures for the excitation of these different radiation patterns.

6.2.2 Feeding Designs for the Excitation of the Synthesized Currents

To demonstrate the novel design concept, compact feed probes are designed and mounted on the UAV platform to excite the ideally synthesized currents. This section will experimentally demonstrate the concept and verify the main electromagnetic features by fabrication and measurement of the scaled UAV prototypes. Moreover, a feed network is designed to control the magnitude and phase excitations for each port, and thus reconfigurable radiation patterns can be obtained from the UAV antenna system.

6.2.2.1 *Geometry and Placement of Feed Probes* As reported in Ref. [9], the optimal locations for probes to excite the currents on its supporting structure are at the locations with maximum currents. As can be observed from Figure 6.6, the current distributions have clearly illustrated the optimal locations, that is, tail and front edges of the two wings. Theoretically, any sources (such as ideal voltage sources or ideal current sheet sources) placed at these locations can excite the synthesized currents and then produce the synthesized radiation patterns as shown in Figure 6.6. However, electrically small radiation systems always suffer from low resistance and substantially large reactance in the input impedance. Input impedance matching is thus very difficult. On the other hand, slot monopoles [23] and inverted F-shaped patch [24] have found wide applications in small wireless terminal devices. Therefore, two slot monopoles and an inverted F-shaped probe are designed and mounted on the front edges of the wings and the tail, respectively. Figure 6.7 shows the theoretical model and the prototype of the three-port UAV-integrated antenna system.

Figure 6.8 shows the geometry of the isolated slot monopole and inverted F-shaped probes. The slot monopole is printed on a 0.8 mm thick FR4 substrate ($\varepsilon_r = 4.4$) with a size of 65 mm × 10 mm. The length of the bended slot is close to a quarter-wavelength at 800 MHz, and a microstrip feed line is printed on the opposite side to excite the slot monopole. The ground plane of the monopole slot is shorted to the front edges of the UAV wings to enhance radiation resistance. The spiral slit and L-shaped stub at the end of the microstrip line are carefully designed for the purpose of input impedance matching.

In addition, planar inverted F-shaped antennas are widely adopted in mobile phones, and the ground plane of mobile phones is generally utilized as part of the radiator. Therefore, the inverted F-shaped probe shown in Figure 6.8b is designed to excite the currents at the tail of the UAV body. It is printed on a 1.6 mm thick FR4 substrate ($\varepsilon_r = 4.4$) with a size of 80 mm × 10 mm. The feed is placed between the open and shortened end of the probe. Its position can be tuned to get good input impedance matching.

Dimensions of the two kinds of probes are turntable for achieving satisfactory impedance matching performance. In the slot monopole probe, increasing the bended slot length will increase the inductance component in the input impedance, whereas

(a)

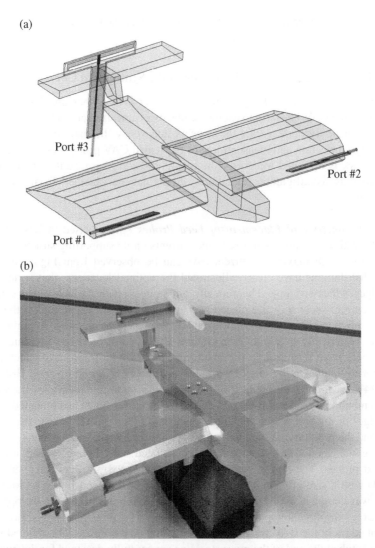

(b)

FIGURE 6.7 Theoretical model (a) and prototype (b) of the three-port UAV integrated antenna system. From Ref. [18]. © 2014 by IEEE. Reproduced by permission of IEEE.

increasing the arm length of the F-shaped probe will increase the capacitance component in the input impedance. The bended slot length and F-shaped arm length can be further reduced in case more miniature probes are demanded. In this case, the input impedance matching can be realized by introducing a matching network inside the UAV body.

It should be noted that, as shown in Figure 6.13, the reflection coefficients for the probes used as isolated antennas are close to 0 dB. It indicates they cannot work independently as stand-alone antennas. To successfully excite the ideally synthesized

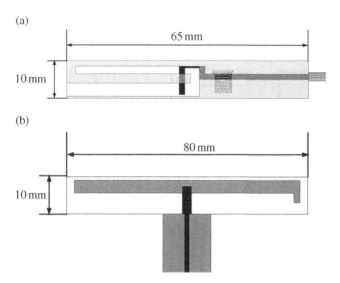

FIGURE 6.8   Geometry of the isolated slot monopole and inverted-F shaped probes. (a) Slot monopole and (b) inverted F-shaped probe.

currents on the UAV body, the coupling from the UAV platform is intentionally enhanced by placing the probes at current maximums. The integration of the UAV platform with three probes potentially yields a reconfigurable radiation system. For communication from different angles, the resultant radiation system can be driven by one or multiple sources.

6.2.2.2   Feeding Network for the Reconfigurable Radiation Patterns   The feed network design is easy and straightforward. Figure 6.9 shows the feed network and photo of the fabricated prototype. The one-to-three Wilkinson power divider directly divides the electromagnetic waves into three output ports with the same energy. Phase delay lines that will achieve 120 and 180° phase delays in the output port #3 and #4 are introduced in their traces.

All the full-wave simulations in this section are conducted using the commercial software Ansoft High Frequency Structure Simulator (HFSS). Figure 6.10 displays the transmission and reflection characteristics to each port. It can be clearly observed that the power divider provides equal power to the output ports, whereas the phase difference between port #2 and port #3 is 120°, and the phase difference between port #2 and port #4 is 180° at 800 MHz. The slight differences in the output power for port #2, #3, and #4 are mainly attributed to different insertion losses introduced by different phase delay lines.

6.2.2.3   Simulated Performance of the Radiation System   When the probes mounted on the UAV body are fed through the output ports of the feed network, the excited current distributions on the UAV body as shown in Figure 6.11 exhibit good approximations to those synthesized currents. Therefore, it is possible to realize three

(a)

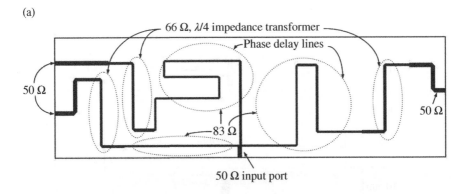

(b)

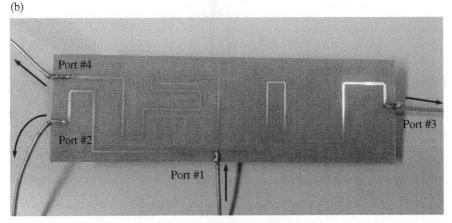

FIGURE 6.9 Theoretical model (a) and photo (b) of the fabricated feed network. From Ref. [18]. © 2014 by IEEE. Reproduced by permission of IEEE.

reconfigurable radiation patterns by properly controlling the magnitude and phase excitations. Figure 6.12 shows the simulated radiation patterns radiated by the excited currents on the UAV platform. Reconfigurable radiation patterns pointing to the broadside, frontward, and backward are obtained.

It should be mentioned that if antennas with size comparable to wavelength are designed first and installed on the platform for sectorial pattern coverage, the coexistence of several antennas requires large space and will also produce undesired interactions between antennas and platforms. Because of the large wavelength in the HF range, these issues will become particularly critical. The present design method promises to provide a suitable solution to deal with these issues.

6.2.3 Experimental Validations

Experimental validation has been carried out for the UAV platform antenna design. Figure 6.13 shows the simulated and measured reflection coefficients of the three-port UAV antenna system. It is observed that the simulated impedance

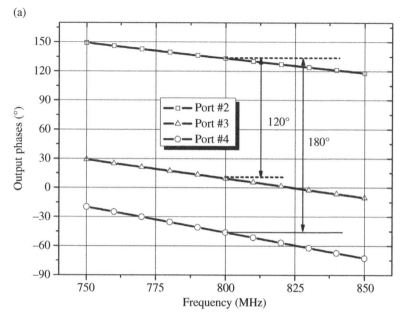

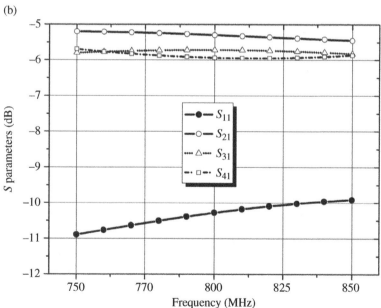

FIGURE 6.10   Transmission and reflection characteristics to each of the ports of the power divider: (a) output phases and (b) transmission and reflection coefficients.

bandwidth is about 2.5% for VSWR (Voltage Standing Wave Ratio) less than or equal to 2. The frequency bands are centered at 800 MHz. In addition, the measured frequency bands are centered at 820 MHz. The impedance bandwidths for the slot monopole probes and inverted F-shaped probe are 2.5 and 1.5% (VSWR ≤ 2), respectively.

Figure 6.14 presents the measured three-dimensional realized gain patterns. The directions of the main beams are in very good agreements with the design objectives.

(a)

(b)

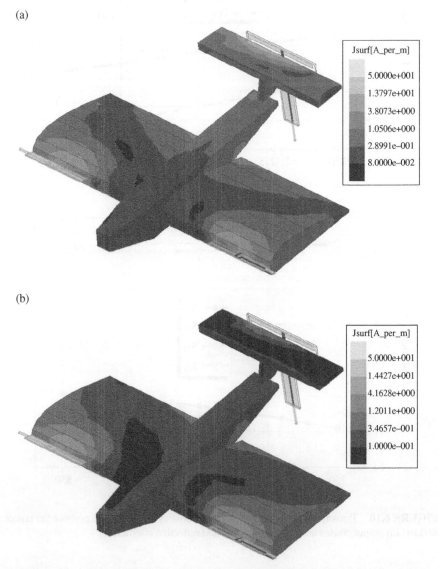

FIGURE 6.11 HFSS simulated current distributions on the UAV platform for (a) broadside, (b) frontward.

(c)

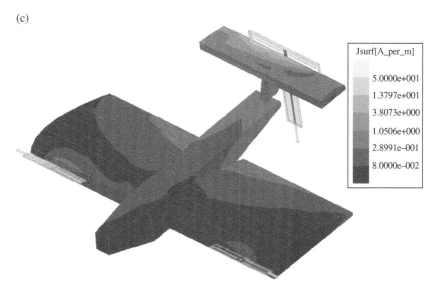

Jsurf[A_per_m]

5.0000e+001
1.3797e+001
3.8073e+000
1.0506e+000
2.8991e−001
8.0000e−002

FIGURE 6.11   (*Continued*) (c) backward radiation cases.

Because of the blockage of the rotational positioner in the chamber, half of the radiation sphere is missing. By eliminating the insertion loss of the feed network, the maximum total realized gains achieved for the broadside, frontward, and backward beams are 3.31, 0.14, and 5.07 dBi, respectively. The peak gain for the frontward radiation pattern is lower than the other two beams, due to its fatter radiation patterns. It should be emphasized that as compared with the HFSS-simulated radiation patterns, the measured radiation patterns for the broadside, frontward, and backward cases agree well with the simulated results. The slight differences may be caused by the feed network, semi-rigid coaxial cables, and supporting structure for the UAV in the far-field measurement. It is difficult to consider these structures altogether accurately in the simulation stage. In summary, both the simulated and measured results have demonstrated the design concept as well as its flexibility for reconfigurable radiation system designs.

6.2.4   Summary

A systematic CM-based approach is developed and demonstrated in an electrically small UAV antenna system design. The pattern synthesis algorithm and operation principles have been discussed with the experimental validation. The realization of a UAV antenna excited by explicitly placed feed probes provides a very helpful hint to achieve good impedance matching (~2% for VSWR ≤ 2) and reconfigurable radiation performance. Three points can be observed from the successful demonstration of the proposed design method. First, taking the electrically small platform as the radiating aperture makes the probes match well with the standard 50 Ω impedance. This concept avoids mismatching problems caused by the low resistance and large

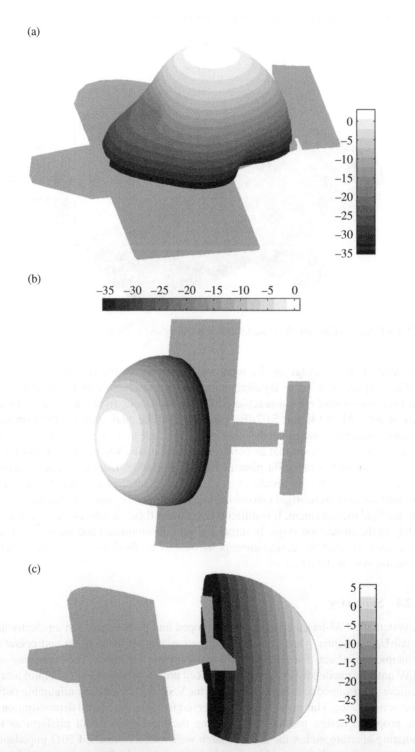

FIGURE 6.12 HFSS-simulated radiation patterns radiated by the excited currents on the UAV platform for (a) broadside, (b) frontward, and (c) backward radiation cases.

(a)

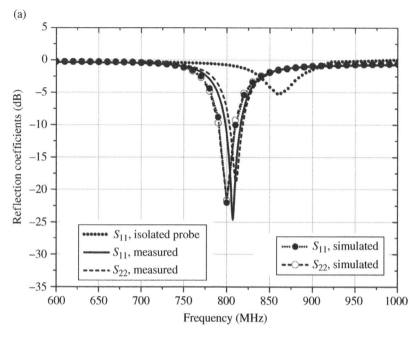

(b)

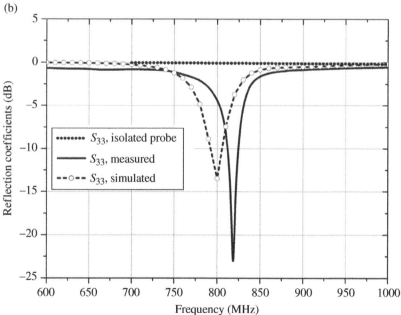

FIGURE 6.13 Simulated and measured reflection coefficients of the three-port UAV antenna system. (a) The slot monopole probe and (b) the inverted F-shaped probe.

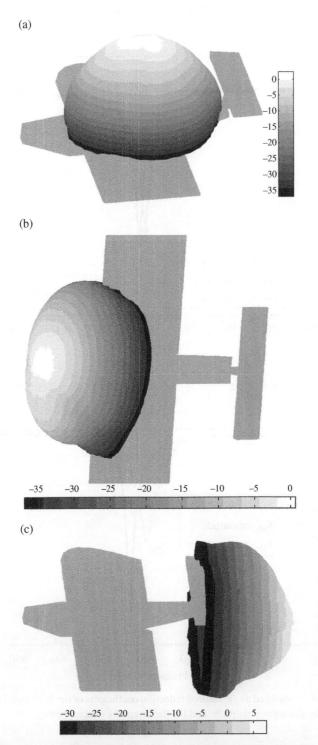

FIGURE 6.14 The three-dimensional measurement of the total realized gain of the UAV antenna system: (a) broadside, (b) frontward, and (c) backward radiation.

reactance that are always faced in conventional electrically small antenna designs. Second, adopting compact and low-profile probes for excitation usages allows the integration of many probes for multifunctional antenna system design. Third, time-consuming full-wave simulations are replaced with linear combinations of the CM fields. The pattern synthesis efficiency is improved up to several orders.

6.3 HF BAND SHIPBOARD ANTENNA SYSTEM DESIGN USING CHARACTERISTIC MODES

Antennas operated at HF band (3–30 MHz) are extremely popular and are indispensable in a plenty of modern shipboard communication systems [25–28]. Take the sky-wave communication as an example; shipboard antennas with broadside radiation patterns center at 5 MHz are required to ensure the radio waves being reflected and returned to the earth by the ionosphere [12, 13]. Because of the large wavelength in this frequency band, the resonant length for efficient radiation will be quite large. Moreover, the size of the entire ship platform or the superstructure on the ship is comparable to the wavelength. As a result, the ship platform falls in the near-field zone of the HF antennas. It leads to many critical issues in practical designs [25]. In particular, antenna performance may degrade due to the strong electromagnetic coupling from the platform. Despite the very limited installation space, antenna engineers have to install large HF antenna at suitable position for least performance deterioration. In addition, conventional HF antennas are usually in the form of wire antennas. These long wire antennas are also easy to be damaged in storms and thunders.

In Section 6.2, we have described the technical details of the CM-based approach for HF band UAV antenna system designs. It is evident that the CMs of the platform are the foundation stone of the approach. With the structural antenna concept, the radiation efficiency of the entire radiation system is enhanced by exciting the platform with properly designed feeding elements. The effectiveness of the approach is demonstrated in an electrically small HF UAV antenna system design.

On the other hand, although the feeding elements used to excite the UAV platform are of low profile and rather compact as compared with the size of the platform, they are the additional protruding structures around the UAV platform. In high-speed moving vehicle platforms, these protruding structures may deteriorate the aerody-namic performance of the original platform. Therefore, protruding feeding structures are not suitable in such circumstances. Conformal or platform-embedded feeding structures are thus highly demanded.

This section continues to discuss the CM-based approach for HF band platform-integrated antenna system designs and explains in full detail the former CM-based structural antenna concept in the following aspects:

- To constrain the dominant radiating currents in a small area of the reserved space of the antenna system, a new CM-based radiation pattern synthesis procedure is proposed. This approach applies the structural antenna concept to platforms with larger size than the resonant length at HF band. It offers new freedom to locate the feedings in the feasible installation area and avoids the intensive currents induced in forbidden areas such as the firing zones. Moreover, the resulting feedings will be close to each other, which will then facilitate the feeding network design. With this approach, the CM-based structural antenna concept becomes more engineering-oriented.

- To make the synthesis procedure more efficient, a criterion is proposed to determine how many significant CMs should be considered in the radiating current synthesis. The contribution of each CM to the synthesized radiation pattern is also investigated to clearly understand how the desired radiation pattern is realized through the combination of CMs.

- To make the HF antenna system fully conformal with the platform, non-protruding slits are proposed using inductive coupling elements to excite the synthesized currents. Good impedance matching is readily obtained by feeding the slits with standard 50 Ω RF coaxial cables.

The newly developed CM-based structural antenna concept is demonstrated through a HF band shipboard antenna system design. The performance of the designed antenna system is experimentally validated in a 1:400 scale ship model [26].

6.3.1 CM-Based Broadside Radiation Pattern Synthesis

To illustrate the design methodology of the newly developed CM-based structural antenna concept, a typical naval ship geometry as shown in Figure 6.15 is considered. We assume shipboard sky-wave communication is required to go beyond an obstacle by taking advantage of the ionospheric reflection in the 5 MHz band. The electrical

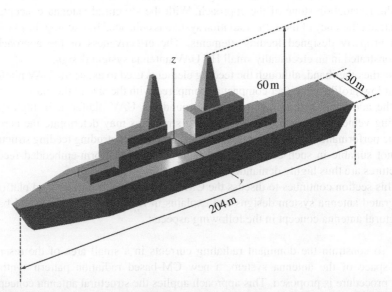

FIGURE 6.15 Geometry and dimensions of the naval ship. From Ref. [26]. © 2015 by IEEE. Reproduced by permission of IEEE.

size of the naval ship at 5 MHz is $3.4\lambda \times 0.5\lambda \times 1.0\lambda$. Because it is larger than the conventional half-wavelength resonant length, only a part of the existing ship structure is sufficient to excite for efficient radiation.

In this design, we assume only the areas on the superstructure are allowed for antenna designs. In sky-wave communications, radiation patterns with high take-off angles are preferred. Therefore, broadside radiation pattern with its radiation power peak toward the zenith is synthesized from the CMs of the ship platform. The following subsections describe the details on the design of platform embedded antennas with broadside radiation pattern using the superstructure of the ship in the 5 MHz band.

6.3.1.1 *Characteristic Modes of the Ship Platform*   For convenience, the definition for the modal significance is rewritten as follows:

$$MS = \left| \frac{1}{1 + j\lambda_n} \right| \tag{6.6}$$

It measures the potential contribution of each mode when an external source is applied to the platform. Moreover, the half-power modal significance is defined to identify the significant modes and non-significant modes,

$$
\begin{cases}
MS \geq \dfrac{1}{\sqrt{2}}, & \text{significant mode} \\[2mm]
MS < \dfrac{1}{\sqrt{2}}, & \text{non-significant mode}
\end{cases}
\tag{6.7}
$$

In the following, this criterion is applied to determine how many significant modes are to be considered in the radiating current synthesis. Following the modal solutions in the CM theory, we again implement the multiobjective optimizer, MOEA/D, to find a set of modal expansion coefficients a_n such that the resulting far fields satisfy the designated radiation pattern.

To understand more about the CMs of the ship platform, see Figure 6.16 that shows the characteristic currents $\mathbf{J}_n$ and characteristic fields $\mathbf{E}_n$ of the first four modes at 5 MHz. As can be observed, the CMs provide clear physical insight into the electromagnetic resonant behavior of the ship platform. In particular, the superstructure dominates the resonance of the first mode $\mathbf{J}_1$. As the order of the modes gets higher, resonant currents also appear on the deck. We can also observe that $\mathbf{J}_1$ and $\mathbf{J}_4$ modes have strong radiation components toward the zenith. However, none of the characteristic fields is the standard broadside radiation patterns as expected. Therefore, we cannot simply excite a single CM to get the designated broadside radiation pattern. Instead, we have to combine these modes properly to get the designated broadside radiation pattern.

Figure 6.17 plots the modal significance of the first 50 modes. It should be noted that the smooth changes of the modal significance with the frequency ensures that the synthesized radiation patterns are stable within a reasonable frequency band. It is observed from Figure 6.17 that there are 20 modes having modal significances larger

(a)

(b)

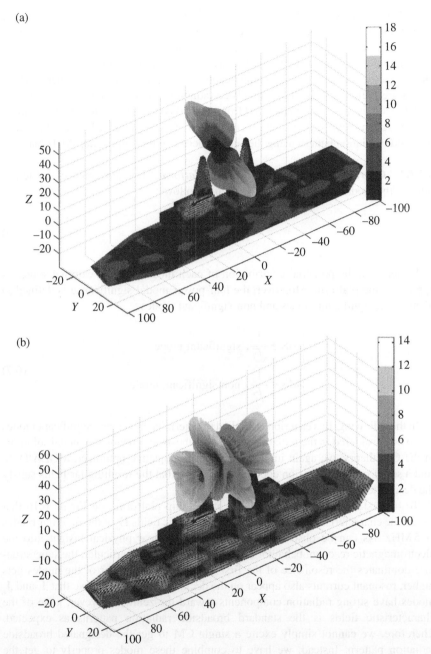

FIGURE 6.16 The first four characteristic currents and fields at 5 MHz. All the color scales are in dB. (a) $\mathbf{J}_1$ & $\mathbf{E}_1$, (b) $\mathbf{J}_2$ & $\mathbf{E}_2$. From Ref. [26]. © 2015 by IEEE. Reproduced by permission of IEEE.

(c)

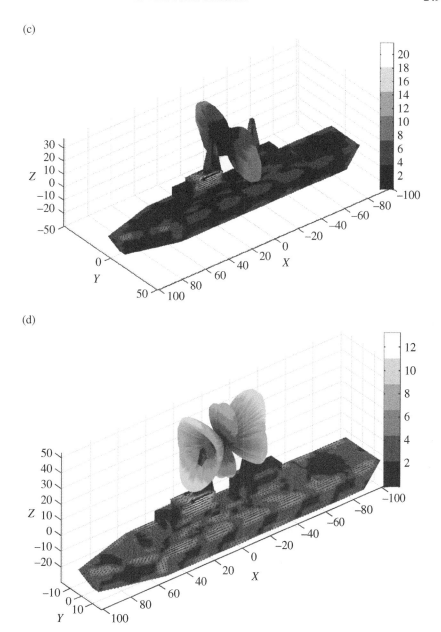

(d)

FIGURE 6.16   (*Continued*) (c) $\mathbf{J}_3$ & $\mathbf{E}_3$, and (d) $\mathbf{J}_4$ & $\mathbf{E}_4$.

than $1/\sqrt{2}$ around 5 MHz. By following the half-power modal significance criterion in Equation (6.7), it should be sufficient to choose these 20 modes as significant modes to do pattern synthesis around 5 MHz, as those non-significant modes contribute little to the final synthesized radiation patterns.

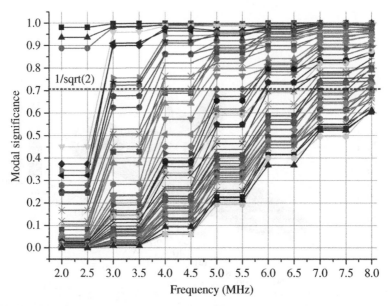

FIGURE 6.17   Modal significances of the first 50 modes. There are 20 significant modes around 5 MHz. From Ref. [26]. © 2015 by IEEE. Reproduced by permission of IEEE.

6.3.1.2   A Modified Radiation Pattern Synthesis Procedure

The radiation pattern synthesis is essentially a multiobjective optimization problem. In this section, the efficient MOEA/D [21, 22] is again combined with the CM theory to synthesize the currents on the ship platform for achieving the designated broadside radiation pattern. Figure 6.18 shows a flowchart that completely illustrates the CM-based synthesis procedure for platform-antenna integration problems. The key steps in this flowchart include the following:

- Solve the CMs of the entire ship platform;
- Compute the modal significance for each mode and pick out the significant modes using the half-power modal significance criterion;
- Define the cost functions for designated radiation pattern and run the optimization to find the modal coefficients a_n by using the modal solutions in Equations (2.76) and (2.77);
- Following specific design objectives, choose the solution from the Pareto front of the MOEA/D optimization and determine the optimal feeding location according to the current distribution;
- Design practical feedings to excite the synthesized radiating currents.

It is observed from Figure 6.16 that the currents on the deck may also contribute to the broadside radiation pattern. However, as stated previously, we assume the antenna designs are only allowed on the superstructure of the ship.

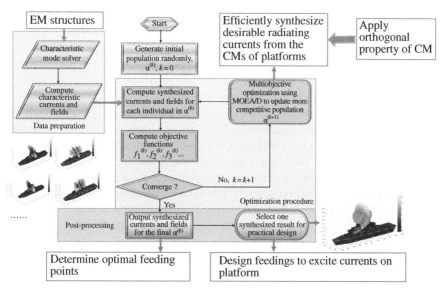

FIGURE 6.18 Flowchart of the CM-based synthesis procedure for platform-antenna integration problems.

To constrain the dominant radiating currents on the superstructure, we construct the following cost functions,

$$f_1\left(\boldsymbol{\alpha}^{(k)}\right) = \left|\theta_0 - \theta_{\text{desired}}\right| + \left|\phi_0 - \phi_{\text{desired}}\right| \tag{6.8}$$

$$f_2\left(\boldsymbol{\alpha}^{(k)}\right) = -\frac{\oiint\limits_{S_{\text{MR}}}\left(\mathbf{E}^* \times \mathbf{H}\right) \cdot d\mathbf{s}}{\oiint\limits_{S_{\infty}}\left(\mathbf{E}^* \times \mathbf{H}\right) \cdot d\mathbf{s}} \tag{6.9}$$

$$f_3\left(\boldsymbol{\alpha}^{(k)}\right) = -\frac{\sum\limits_{i \in \text{Super}}\left|J_i \mathbf{f}_i\right|}{\sum\limits_{i=1}^{N}\left|J_i \mathbf{f}_i\right|} \tag{6.10}$$

The cost function defined in Equation (6.8) specifies the main beam direction of the radiation pattern, where θ_0 and φ_0 are the calculated main beam directions corresponding to a particular set of modal coefficients $\boldsymbol{\alpha}^{(k)} = \left\{a_1^{(k)}, a_2^{(k)}, \ldots, a_{20}^{(k)}\right\}$ at the kth iteration in the multiobjective MOEA/D optimizer. Because 20 significant CMs are found around 5 MHz, the modal coefficients $\boldsymbol{\alpha}^{(k)}$ comprise 20 complex coefficients. The decision spaces for the amplitude and phase of $a_i^{(k)}$ are [0, 1] and $[-\pi, \pi]$, respectively. To achieve the designated radiation pattern with its peak toward the zenith, we apply the desired main beam direction $\theta_{\text{desired}} = \phi_{\text{desired}} = 0°$ in Equation (6.8).

In the second cost function defined in Equation (6.9), $f_2(\boldsymbol{\alpha}^{(k)})$ specifies the ratio between the radiated power over the main beam range S_{MR} and the entire radiation sphere S_∞. In the present example, S_{MR} is defined as the cone range of $0° \leq \theta \leq 45°$ and $0° \leq \varphi \leq 360°$. Because the MOEA/D algorithm is developed to minimize the value of the cost function, a negative sign is assigned to ensure that the antenna system radiates more energy in the main beam range.

The last cost function distinguishes the present radiation pattern synthesis procedure from the one in Equation (6.5). In this cost function, J_i is the complex weighting coefficients for the RWG (Rao-Wilton-Glisson) basis functions defined on the naval ship structure. Assuming N basis functions $\mathbf{f}_i$ are defined on the surface of the entire ship structure, the cost function in Equation (6.8) calculates the ratio of the current intensity on the superstructure to that on the entire ship platform. $i \in$ Super denotes the basis functions that are assigned on the superstructure. Similarly, a negative sign is also assigned. It ensures the dominant radiating currents be localized on the superstructure.

Owing to the modal solutions of the CM theory, the far fields for any set of $\boldsymbol{\alpha}^{(k)}$ can be calculated efficiently through a superposition of the CMs and the accuracy is the same as in the full-wave simulations. The benefit is that the modal solution helps to achieve a speedup up to three orders in the optimization stage [18, 29, 30]. The cost function values in Equations (6.8)–(6.10) can be readily obtained once the far fields have been calculated.

6.3.1.3 *Synthesized Radiation Pattern* Applying the cost functions in Equations (6.8)–(6.10) in the pattern synthesis procedure, we get a Pareto front of the multiobjective optimization problem. The Pareto front is an objective space that comprises many tradeoff solutions. In the Pareto front, no solution is better than the other one. We can select one solution from the Pareto front according to specific design objectives.

Figure 6.19 shows the amplitudes and phases of the modal coefficients $\boldsymbol{\alpha}^{(k)}$ of our selected solution in the Pareto front. This set of modal coefficients gives a satisfactory radiation pattern that approximates well to the designated radiation pattern. Using the same set of $\boldsymbol{\alpha}^{(k)}$, we can directly calculate the corresponding radiating currents using Equation (2.75). A careful investigation to the Figure 6.19 reveals that the modes $\mathbf{J}_1$ and $\mathbf{J}_4$ are weighted with larger modal coefficients than the modes $\mathbf{J}_2$ and $\mathbf{J}_3$. On the other hand, Figure 6.16 shows that the modes $\mathbf{J}_1$ and $\mathbf{J}_4$ have larger component in the zenith direction than the modes $\mathbf{J}_2$ and $\mathbf{J}_3$. Similar phenomena can be observed for the other 16 modes. Evidently, the pattern synthesis procedure has efficiently enhanced the contribution from desired modes and suppressed the contribution from unwanted modes.

To figure out how to excite the synthesized currents, let us take a look at the synthesized radiation pattern and the radiating currents as shown in Figure 6.20. As can be seen, the radiation pattern is a standard broadside radiation pattern. The main beam is in the direction of the zenith, and the sidelobe levels are in acceptable level. Moreover, the dominant radiating currents are constrained on the lower parts of the

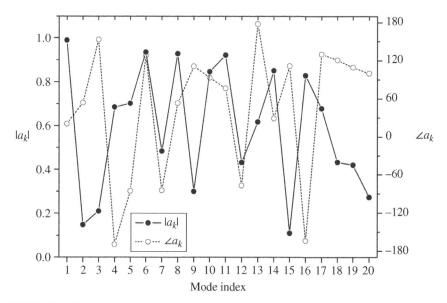

FIGURE 6.19 Modal expansion coefficients for the synthesized currents and broadside radiation pattern.

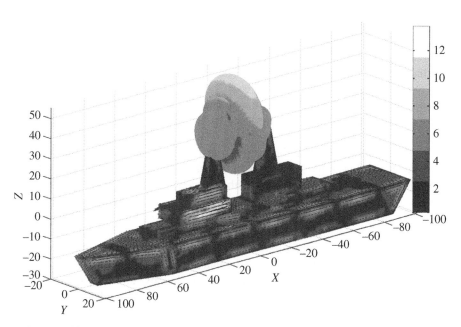

FIGURE 6.20 The synthesized currents and broadside radiation pattern. The color scale is in dB. From Ref. [26]. © 2015 by IEEE. Reproduced by permission of IEEE.

superstructure. Although the currents on the deck also contribute to the synthesized pattern, they are much weaker than those on the superstructure. Therefore, to simplify the feeding structure design, it is reasonable to ignore these currents.

6.3.2   Feeding Structure Design

It has been long understood that tilted whips or monopoles are probably the only candidates for sky-wave communications [12, 31, 32]. To increase the radiation power at high take-off angles, these wire antennas are usually loaded with various lump elements. However, whips or monopoles suffer from low radiation power in a ±45° cone range centered in the zenith [32]. As a result, loop antennas, which offer improved performance due to the existence of a significant horizontal current component for radiation toward the zenith, are preferred [32]. However, loop antennas are difficult to be fully conformal with the platform or embedded into the platform. Therefore, these wire antennas are not considered in this work.

On the other hand, an earlier work of Newman enlightens us on the issue of how to excite the radiating currents on platforms [9]. Newman suggested that the optimum location of a small loop that excited the supporting structure should be at the position with maximum currents. As suggested by Martens et al. [33], the coupling elements used to excite currents on platforms can be categorized into inductive coupling element (ICE) and capacitive coupling element (CCE). It was demonstrated that the ICE placed at the current maxima and the CCE placed at the current minima can excite a specific current independently. It was also showed that the ICE offers advantages over CCE in terms of the mode purity. A very recent paper [34] proposed a hybrid coupling element to get wider impedance bandwidth.

In this design, we also aim to design feeding structures being able to embed into the ship platform. As the CCE usually introduces an additional element to the original structures, it is not suitable to the present design. Alternatively, the ICE usually appears as a slot cut on the original structure and fed with a voltage source [33–35]. There is no additional element introduced to the original structure. Therefore, we adopt the ICE scheme in this work.

As shown in Figure 6.20, the synthesized radiating currents have the current maxima on the lower parts of the superstructure. Therefore, the ICEs will be implemented by introducing two slits with voltage source on the lower parts of the superstructure. To obtain a prototype with reasonable size for ease of measurement in the microwave anechoic chamber, a shipboard antenna system is designed through a 1:400 scale ship model. The 5 MHz band is thus scaled to the 2.0 GHz band.

Figure 6.21 shows the configurations of the ICE feedings for the excitation of the synthesized currents with providing coordinate system for the completeness of understanding. Standard 50 Ω RF coaxial cables are directly connected to the slits. To provide a delta voltage source for the slits, the center conductor and outer conductor of the coaxial cables are soldered to the two edges of the slits, respectively.

On the other hand, the synthesized radiating currents show that currents on the two sides of the superstructure have the same current magnitude and 180° phase difference. Therefore, a Wilkinson power divider is designed to feed the two slits

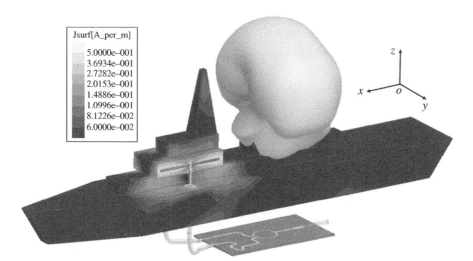

FIGURE 6.21 The excited current distributions on the scaled ship platform and radiating far fields at 2.08 GHz. ICEs are used to excite the synthesized currents. From Ref. [26]. © 2015 by IEEE. Reproduced by permission of IEEE.

with equal power and 180° phase difference. To ensure the power balance and high isolation between two output ports, a 100 Ω resistor is introduced into the Wilkinson power divider (as shown in Fig. 6.25). In addition, a phase delay line is introduced on the way to port 2. The two output ports thus have a 180° phase difference. The impedance transformer on the way to input port transforms the characteristic impedance of the microstrip line to the standard 50 Ω. The characteristics of the power divider are given in Figure 6.22. As can be seen, because of the loss of the dielectric material and copper conductor, a 1 dB insert loss is introduced on the two output ports. Figure 6.23 shows the simulated return loss of the 1:400 scaled antenna system compared with measured results. It shows that the system matches well to the standard 50 Ω system with a reasonable bandwidth.

The full-wave simulated currents and three-dimensional radiation pattern excited by the designed feeding structures are shown in Figure 6.21. As can be seen, they approximate the theoretically synthesized currents and pattern in a reasonable sense.

Figure 6.24 shows the simulated radiation patterns at the lower, center, and upper frequencies, respectively. According to the coordinate system defined in the Figure 6.21, the *xoz* and *yoz* planes are the H-plane and E-plane, respectively. As can be seen, the radiation patterns are stable across the frequency band. The main beam is in the direction of the zenith, which meets the requirement in the sky-wave communications. Taking into account the insert loss introduced by the power divider and coaxial cables, the realized gain is about 1.7 dBi at the center frequency. However, to achieve the same gain using wire antennas such as the loop [36] or whip [37] antennas, the antenna size will be extremely large, and sufficient installation height ($\approx \lambda/4$) over the platform/ground plane is also necessary, which is always impractical.

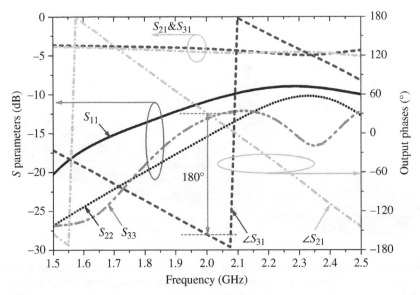

FIGURE 6.22 The transmission and port characteristics of the power divider. From Ref. [26]. © 2015 by IEEE. Reproduced by permission of IEEE.

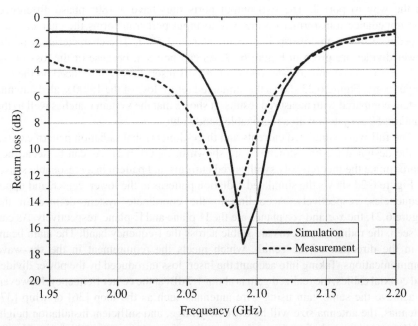

FIGURE 6.23 Simulated and measured return loss of the 1:400 scale model. From Ref. [26]. © 2015 by IEEE. Reproduced by permission of IEEE.

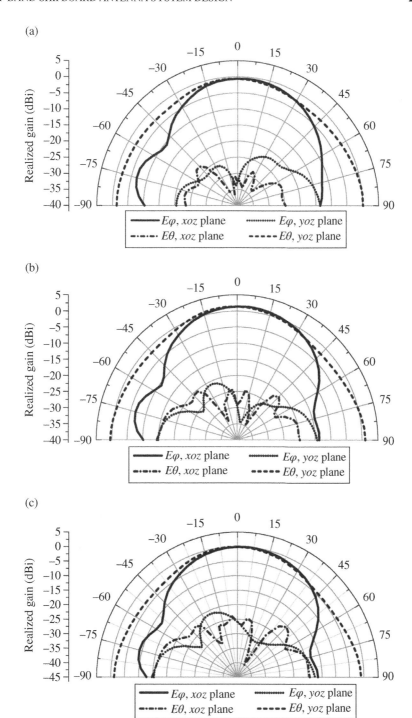

FIGURE 6.24 The simulated radiation pattern. (a) 2.05 GHz, (b) 2.08 GHz, and (c) 2.12 GHz. From Ref. [26]. © 2015 by IEEE. Reproduced by permission of IEEE.

6.3.3 Experimental Validations

To validate the radiation performances of the scale model, a prototype is fabricated and measured. Figure 6.25 shows the assembled radiation antenna system. The main body of the ship model is made of aluminum, while the lower part of the tower is made of brass for ease of soldering purpose. The measured return loss of the scale model is compared with the simulated results in Figure 6.23. It is observed that the design achieves reasonable bandwidth and also has good agreement with the simulated return loss.

In the far-field measurement setup, the ship platform and its supporting structure, as shown in Figure 6.25, are mounted over a conducting plane. The radiation patterns measured in this measurement setup are shown in Figure 6.26a. As can be seen, they are standard broadside radiation patterns with their maximum radiations in the zenith. Figure 6.26b shows the simulated radiation patterns with the similar measurement environment. They are in reasonable agreement with the measured radiation patterns. A realized gain of about 1.7 dBi is also obtained at the center frequency, which agrees with the simulated gain. The cross polarization level is less than 20 dB in each plane. The measured realized gain and radiation efficiency of the shipboard antenna is given in Figure 6.27. As can be seen, the realized gain is stable across the frequency band. The radiation efficiency around the center frequency is larger than 40%, which is quite good as compared to the conventional platform-installed HF wire antennas.

As have been demonstrated, by properly designing the feeding structures and exciting them with proper RF signals, the radiating currents synthesized from the

FIGURE 6.25 The fabricated prototype of the 1:400 scale model. From Ref. [26]. © 2015 by IEEE. Reproduced by permission of IEEE.

ship platform's CMs are successfully excited. The following points should be highlighted from the experimental validation:

- The radiating currents synthesized from the CMs can be practically excited.
- The synthesized currents provide useful information for the feeding structure placement.
- By taking advantage of the large electrical size of the superstructure on the platform, satisfactory performance such as the realized gain or the radiation efficiency of the entire antenna system is guaranteed.
- Finally, yet importantly, protruding structures are not necessary, which makes the radiation system fully conformal with the platform.

(a)

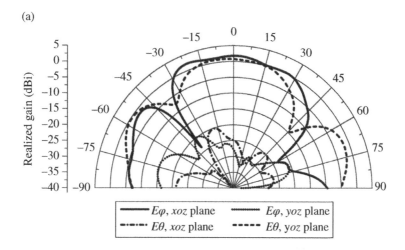

(b)

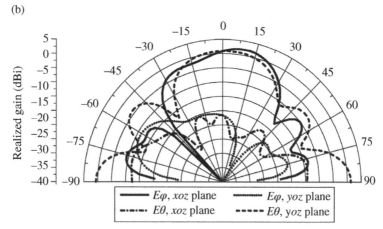

FIGURE 6.26 Measured (a) and simulated (b) radiation patterns at 2.08 GHz. From Ref. [26]. © 2015 by IEEE. Reproduced by permission of IEEE.

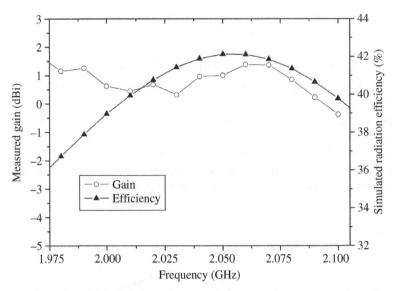

FIGURE 6.27 Measured realized gain and simulated efficiency of the shipboard antenna. From Ref. [26]. © 2015 by IEEE. Reproduced by permission of IEEE.

6.3.4 Practical Implementation

After conducting the theoretical and experimental study on the CM-based shipboard antenna designs, this section will continue to validate the proposed design concept through a practical implementation on a realistic ship in the 5 MHz band. Because of the large size of the ship and large wavelength in the 5 MHz band, conducting measurement is impossible. Therefore, this section will focus on the theoretical validation through accurate full-wave simulation using the commercial software ANSYS HFSS.

One can observe that owing to the limited space inside the 1:400 scaled model (as shown in Fig. 6.25), the RF cables and power divider are assembled out of the ship for easy soldering. In practical implementation, however, all the feeding structures and feeding networks could be assembled inside the realistic ship. Figure 6.28 shows an explosive view of the practical implementation of the design. As can be seen, the feeding cables are assembled inside the ship, and no exterior protruding element can be observed. Therefore, the design meets the critical demands of conformal or platform-embedded HF antenna designs. In the same way, the center and outer conductor of the coaxial cables are connected to the two edges of the slit to provide a voltage source. However, it should be pointed out that the Wilkinson power divider shown in Figure 6.25 is only for experimental validation in laboratory. In practical implementation, commercial standard power divider and phase shifter are available and can be incorporated inside the ship to offer equal power and 180° phase differences.

Figure 6.29 shows the simulated active reflection coefficients of the practical design in HF band, where Port 1 and Port 2 refer to the two coaxial cable feeding

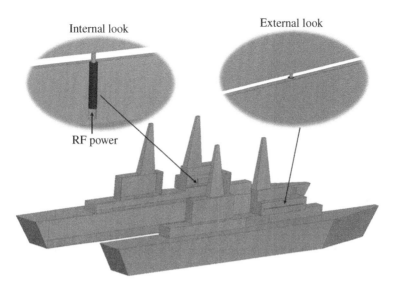

FIGURE 6.28 Explosive view of the practical implementation of the design in HF band. From Ref. [26]. © 2015 by IEEE. Reproduced by permission of IEEE.

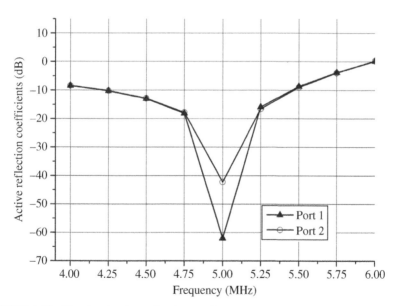

FIGURE 6.29 Simulated active reflection coefficients of the practical design in HF band. From Ref. [26]. © 2015 by IEEE. Reproduced by permission of IEEE.

ports for the slits. As can be seen, both ports match well to the standard $50\,\Omega$ system. This ensures good matching to a standard commercial power divider at $5\,\text{MHz}$. Figure 6.30 shows the simulated three-dimensional radiation pattern at $5\,\text{MHz}$. As expected, this is a standard broadside radiation pattern. Good agreement is observed

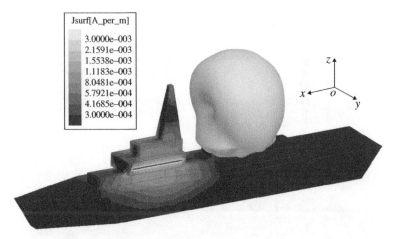

FIGURE 6.30 Simulated radiation pattern and radiation currents of the practical design at 5 MHz. From Ref. [26]. © 2015 by IEEE. Reproduced by permission of IEEE.

as compared to the simulated radiation pattern of the scale model shown in Figure 6.21. It also shows that scaling property in terms of the frequency and physical size is kept between the full and scale models.

In summary, the proposed design approach of using the combination of the CM theory and structural antenna concept has been successfully demonstrated through the theoretical and experimental realization of a platform conformal or platform embedded antenna system on a ship in the HF band.

6.3.5 Summary

A straightforward, easy, and efficient CM-based approach for the design of platform-embedded HF shipboard antenna system has been presented. The CMs of the ship platform are calculated and employed in the synthesis of the desired radiation patterns. Therefore, this approach exploits the physics of the entire platform since the pattern synthesis stage. Moreover, it allows the localization of the synthesized currents on a small part of the existing platform. It makes the CM-based structural antenna concept more suitable to electrically medium or large platforms. As an inductive coupling element, non-protruding slit is adopted to excite the synthesized currents on the platform. The novel CM-based structural antenna concept is then demonstrated through the design of a platform-embedded HF antenna system for sky-wave communications. It is the first time to design platform-embedded HF antenna system. A 1:400 scale antenna system has been fabricated and measured. Measurement results are in good agreement with the full-wave simulation results. Eventually, practical implementation of the design and its resulting performance is examined through accurate full-wave simulations.

6.4 CONCLUSIONS

This chapter has fully discussed the CM-based approach for platform-integrated antenna system designs. Essentially, this CM-based approach tells how to apply the structural antenna concept in practical platform antenna designs. Actually, the structural antenna concept has a long history in real applications. For example, in shipboard communication systems with traditional wire antennas, the ship platform may also contribute a lot to the radiation at some frequencies. In this sense, the wire antenna can be recognized as a feeding to the ship platform. Although the structural antenna concept is believed to be attractive in platform antenna designs, there is no systematic approach existing to show how to make the platform radiating effectively.

The CM-based approach in this chapter gives a solution to the structural antenna concept. It emphasizes that the structural antenna concept can be really implemented in platform-integrated antenna designs. The main merits of this approach are summarized as follows:

- It is applicable to both electrically small and large platforms. In the electrically small case, the entire platform is used as the radiator. In the electrically large case, only part of the platform is used as the radiator.
- In theory, arbitrary radiation patterns can be synthesized from the CMs.
- High efficiency in the radiation pattern synthesis. It is benefited from the modal solution in the CM theory.
- High radiation efficiency or gain. It is benefited from the large radiating aperture: the platform.
- Low profile or even platform embedded feeding structures.
- Owing to the graphic display of the synthesized currents for desired radiation pattern, it is easy to understand how to place feeding structures for maximum power couplings.

REFERENCES

[1] J. V. N. Granger, "Shunt excited flat plate antennas with applications to aircraft structures," *Proc. IRE*, vol. 38, no. 3, pp. 280–287, Mar. 1950.

[2] J. V. Tanner, "Shunt and notch-fed aircraft antennas," *IRE Trans. Antennas Propag.*, vol. 6, no. 1, pp. 35–43, Jan. 1958.

[3] G. Marrocco and P. Tognolatti, "New method for modeling and design of multiconductor airborne antennas," *Proc. IEE Microw. Antennas Propag.*, vol. 151, no. 3, pp. 181–186, Jun. 2004.

[4] K. S. Kunz and R. J. Luebbers, *The Finite Difference Time Domain Method for Electromagnetics* (1st edition), Boca Raton, FL: CRC Press, 1993.

[5] Remcom [online]: http://www.remcom.com/xf7. Accessed January 28, 2015.

[6] J. K. Infantolino, M. J. Barney, and R. Haupt, "Optimal position for an antenna using a genetic algorithm," in IEEE Military Communications Conference, Boston, MA, pp. 1–5, Oct. 2009.

[7] J. K. Infantolino, M. J. Barney, and R. Haupt, "Using a genetic algorithm to determine an optimal position for an antenna mounted on a platform," in IEEE Military Communications Conference, Boston, MA, pp. 1–6, Oct. 2009.

[8] M. J. Barney, J. M. Knapil and R. Haupt, "Determining an optimal antenna placement using a genetic algorithm," in IEEE International Symposium on Antennas and Propagation Society (APSURSI '09), Charleston, SC, pp. 1–4, Jun. 2009.

[9] E. H. Newman, "Small antenna location synthesis using characteristic modes," *IEEE Trans. Antennas Propag.*, vol. AP-21, no. 4, pp. 530–531, Jul. 1979.

[10] K. Obeidat, R. G. Rojas, and B. Raines, "Design of antenna conformal to V-shaped tail of UAV based on the method of characteristic modes," in 3rd European Conference on Antennas and Propagation, Berlin, Germany, pp. 2493–2496, Mar. 2009.

[11] J. Chalas, K. Sertel, and J. L. Volakis, "Design of in-situ antennas using platform characteristic modes," in 2013 IEEE International Symposium on Antennas Propagation, Orlando, FL, pp. 1508–1509, Jul. 2013.

[12] B. A. Austin and K. P. Murray, "The application of characteristic-mode techniques to vehicle-mounted NVIS antennas," *IEEE Antennas Propag. Mag.*, vol. 40, no. 1, pp. 7–21, Feb. 1998.

[13] K. P. Murray and B. A. Austin, "Synthesis of vehicular antenna NVIS radiation patterns using the method of characteristic modes," *IEE Proc.-Microw. Antennas Propag.*, vol. 141, no. 3, pp. 151–154, Jun. 1994.

[14] Y. Chen and C.-F. Wang, "Synthesis of platform integrated antennas for reconfigurable radiation patterns using the theory of characteristic modes," in 10th International Symposium on Antennas, Propagation & EM Theory (ISAPE), Xian, pp. 281–285, Oct. 2012.

[15] Y. Chen and C.-F. Wang, "Shipboard NVIS radiation system design using the theory of characteristic modes," in 2014 IEEE International Symposium on Antennas Propagation, Memphis, TN, pp. 852–853, Jul. 2014.

[16] J. Chalas, K. Sertel, and J. L. Volakis, "NVIS synthesis for electrically small aircraft using characteristic modes," in 2014 IEEE International Symposium on Antennas Propagation, pp. 1431–1432, Jul. 2014.

[17] M. Ignatenko and D. S. Filipovic, "Application of characteristic mode analysis to HF low profile vehicular antennas," in 2014 IEEE International Symposium on Antennas Propagation, Memphis, TN, pp. 850–851, Jul. 2014.

[18] Y. Chen and C.-F. Wang, "Electrically small UAV antenna design using characteristic modes," *IEEE Trans. Antennas Propag.*, vol. 62, no. 2, pp. 535–545, Feb. 2014.

[19] X. Yang, K. Ng, S. Yeung, and K. F. Man, "Jumping genes multiobjective optimization scheme for planar monopole ultrawideband antenna," *IEEE Trans. Antennas Propag.*, vol. 56, no. 12, pp. 659–666, Dec. 2008.

[20] Y. Chen, S. Yang, and Z. Nie, "Improving conflicting specifications of time-modulated antenna arrays by using a multiobjective evolutionary algorithm," *Int. J. Numer. Modell.*, vol. 25, pp. 205–215, May–Jun. 2012.

[21] Q. Zhang and H. Li, "MOEA/D: a multiobjective evolutionary algorithm based on decomposition," *IEEE Trans. Evol. Comput.*, vol. 11, no. 6, pp. 712–731, Jun. 2007.

[22] H. Li and Q. Zhang, "Multiobjective optimization problems with complicated Pareto sets, MOEA/D and NSGA-II," *IEEE Trans. Evol. Comput.*, vol. 13, no. 2, pp. 284–301, Feb. 2009.

[23] K.-L. Wong and W.-J. Lin, "WWAN printed monopole slot antenna with a parallel-resonant slit for tablet computer application", *Microw. Opt. Technol. Lett.*, vol. 55, no. 1, pp. 40–45, Jan. 2013.

[24] J. Lal, H. K. Kan, and W. S. T. Rowe, "Dual-frequency F-shaped shorted patch antenna," *Microw. Opt. Technol. Lett.*, vol. 48, no. 9, pp. 1811–1812, Sep. 2006.

[25] S. R. Best, "On the use of scale brass models in HF shipboard communication antenna design," *IEEE Antennas Propag. Mag.*, vol. 44, no. 2, pp. 12–23, Apr. 2002.

[26] Y. Chen, C.-F. Wang, "HF band shipboard antenna design using characteristic modes" *IEEE Trans. Antennas Propag.*, vol. 63, no. 3, pp. 1004–1013, Mar. 2015.

[27] P. J. Baldwin, A. G. P. Boswell, D. C. Brewster, and J. S. Allwright, "Iterative calculation of ship-borne HF antenna performance," *IEE Proc. H Microw. Antennas Propag.*, vol. 138, no. 2, pp. 151–158, Apr. 1991.

[28] J. Baker, H. S. Youn, N. Celik, and M. F. Iskander, "Low-profile multifrequency HF antenna design for coastal radar applications," *IEEE Antennas Wirel. Propag. Lett.*, vol. 9, pp. 1119–1122, 2010.

[29] Y. Chen and C.-F. Wang, "Synthesis of reactively controlled antenna arrays using characteristic modes and DE algorithm," *IEEE Antennas Wirel. Propag. Lett.*, vol.11, pp. 385–388, 2012.

[30] Y. Chen and C.-F. Wang, "Electrically loaded Yagi-Uda antenna optimizations using characteristic modes and differential evolution," *J. Electromagn. Waves Appl.*, vol. 26, no. 8–9, pp. 1018–1028, Jul. 2012.

[31] B. A. Austin and W.-C. Liu, "Assessment of vehicle-mounted antennas for NVIS applications," *IEE Proc. Microw. Antennas Propag.*, vol. 149, no. 3, pp. 147–152, Jun. 2002.

[32] B. A. Austin and W.-C. Liu, "An optimised vehicular loop antenna for NVIS applications," in 8th International Conference on HF Radio Systems and Techniques, Guildford, pp. 43–47, Jul. 2000.

[33] R. Martens, E. Safin, and D. Manteuffel, "Inductive and capacitive excitation of the characteristic modes of small terminals", in Antennas and Propagation Conference (LAPC), 2011 Loughborough, UK, pp. 1–4, Nov. 2011.

[34] R. Martens and D. Manteuffel, "Mobile LTE-A handset antenna using a hybrid coupling element," in IEEE International Symposium Antennas Propagation, Memphis, TN, pp. 1419–1420, Jul. 6–11, 2014.

[35] D. Poopalaratnam, K. K. Kishor, and S. V. Hum, "Multi-feed chassis-mode antenna with dual-band MIMO operation," in IEEE International Symposium Antennas Propagation, Memphis, TN, pp. 1427–1428, Jul. 6–11, 2014.

[36] M. Koubeissi, B. Pomie, and E. Rochefort, "Perspectives of HF half loop antennas for stealth combat ships," *Prog. Electromagn. Res. B*, vol. 54, pp. 167–184, 2013.

[37] R. P. Milione. "NVIS Antenna Theory and Design," [Online]: http://www.radioscanner. ru/files/antennas/file17195/. Accessed January 28, 2015.

INDEX

Characteristic Modes: Theory and Applications in Antenna Engineering, First Edition.
Yikai Chen and Chao-Fu Wang.
© 2015 John Wiley & Sons, Inc. Published 2015 by John Wiley & Sons, Inc.

Printed and bound by CPI Group (UK) Ltd, Croydon, CR0 4YY

16/04/2025

14658587-0001